A Pluralist Theory of Perception

A Pluralist Theory of Perception

Neil Mehta

The MIT Press
Cambridge, Massachusetts
London, England

The MIT Press would like to thank the anonymous peer reviewers who provided comments on drafts of this book. The generous work of academic experts is essential for establishing the authority and quality of our publications. We acknowledge with gratitude the contributions of these otherwise uncredited readers.

This book was set in Stone Serif and Stone Sans by Westchester Publishing Services. Printed and bound in the United States of America.

Library of Congress Cataloging-in-Publication Data

Names: Mehta, Neil, author.
Title: A pluralist theory of perception / Neil Mehta.
Description: Cambridge, Massachusetts : The MIT Press, [2024] | Includes bibliographical references and index.
Identifiers: LCCN 2023036889 (print) | LCCN 2023036890 (ebook) | ISBN 9780262548281 (paperback) | ISBN 9780262379151 (epub) | ISBN 9780262379144 (pdf)
Subjects: LCSH: Consciousness. | Perception. | Pluralism.
Classification: LCC B808.9 .M374 2024 (print) | LCC B808.9 (ebook) | DDC 121/.34—dc23/eng/20240116
LC record available at https://lccn.loc.gov/2023036889
LC ebook record available at https://lccn.loc.gov/2023036890

10 9 8 7 6 5 4 3 2 1

For Wa Wa, Alison, and Youngmee. Others have made me a better philosopher; you have made me a better person.

Contents

Preface

Perception and consciousness make for a complex knot. Many have tried to undo it, often showing great ingenuity in the process, but I believe that they have left things tied up in the end. In this book, I try to do better. I defend a radical new theory of perception and consciousness (and, for that matter, of hallucination, illusion, and more).

P.1 Methodology; the Core Idea

One natural methodology is to begin with our *commonsense theory* of perception and to assume that we are justified in accepting this theory unless powerful defeaters are found. I think that this methodology is misguided. In my view, a theory enjoys no justification whatsoever simply in virtue of being commonsense. Any theory, commonsense or otherwise, is justified purely by its theoretical virtues—its simplicity, explanatory power, fruitfulness, and so on. Thus, when I was writing this book, I did not start with theories at all. I started with *data* and used these to guide me in assessing old theories and constructing new ones.

I take an ecumenical attitude toward different sources of data: the correct theory, I take it, should explain *all* of the data, whether it comes from armchair introspection or scientific inquiry. That said, my emphasis in this book is firmly on data from armchair introspection.

I take this approach for two reasons. On the one side, my considered judgment is that most of the (non-introspective) scientific data are too low level to discriminate between different high-level philosophical theories of perception: these data can often be given a representationalist, naïve realist, or other interpretation. On the other side, I think that the introspective data identified in the book are more high level and *do* discriminate between different philosophical theories of perception.[1] Ultimately, however, I welcome any empirical arguments that others might make, and I sometimes advance such arguments myself.

As I was writing this book, I began by making a list of familiar data: data pertaining to revelation, the hard problems of consciousness, presentation, knowledge of reference, and much, much more. I then attempted to extend these data as systematically as I could. For example, it is a familiar datum that perception positions us to know, in a special way, what we are referring to. This inspired me to ask whether the same was true of any other kinds of experiences—hallucinations, sensations, sensory imaginings, episodic memories. In this way, I uncovered a great deal of new data that fell into complex and surprising patterns.

Only then did theory enter the picture. I toyed with dozens of theories—old, new, and often radically different from one another—in search of the most elegant possible explanation of these patterns. The result was surprising: the usual representationalist and naïve realist options had a lot of trouble explaining the data. The problem was that these theories, though otherwise disparate, were all *monistic*: they all shared the seemingly benign assumption that to consciously perceive is to deploy exactly one kind of sensory awareness. But this assumption seemed to fit the data poorly.

I realized that by discarding monism, I could construct a new theory with great explanatory power. I call this theory *rich pluralism*. The rich pluralist's signature idea is that what it is to perceive is to deploy two very different kinds of sensory awareness in concert. First, there is a *representational* form of sensory awareness whose targets include *particulars*. This distinguishes perceptions from hallucinations. Second, there is a *nonrepresentational* form of sensory awareness whose targets are *certain universals*—the sensory qualities. This latter form of sensory awareness constitutes one kind of consciousness, it is common to perceptions and hallucinations, and it reveals part of the very essences of its targets. I call it *deep awareness*.

Rich pluralism appears to be less parsimonious than monism. Surprisingly, it is not: the adequate monistic theories, those that can explain all of the data, end up being even more complex than rich pluralism. Thus, the sophisticated apparatus of rich pluralism is not a luxury. It is the bare minimum that we need, the most spartan theory that can do the work. Or so I will argue.

P.2 Acknowledgments

The academy is not like life: we are enriched rather than impoverished by our debts. That is fortunate, for my debts have piled up over the course of this project.

My largest debts are to Brian Cutter and Boyd Millar, whose engagement has vastly improved the book. Brian and Boyd read many manuscripts of articles leading up to the book, gave me detailed comments on multiple complete drafts of the book, and took part in the book workshop. I have to add that Brian's comments on an early draft were, quite simply, the best philosophical comments that I have ever received.

I am also grateful to the other participants in the book workshop—Keith Allen, Heather Logue, and Umrao Sethi—for six hours of heady discussion, and to Keith and Umrao for also having read and commented on substantial portions of an earlier draft. I also thank Ivan Ivanov and Craig French for detailed written comments on the entire manuscript, Alex Moran for written comments on the entire manuscript *and* an illuminating two-hour conversation about the book, and Ori Beck for spending four hours patiently walking me through his concerns on every chapter. All of these individuals kindly revealed much that was previously hidden from my sight.

It will also become apparent to the reader that I have taken much inspiration from the work of Mark Johnston. His 2004 paper "The Obscure Object of Hallucination" is my favorite article on the philosophy of perception. The differences between my theory and his are numerous, but still, they are theories grown from the same seed.

I have received feedback on various pieces of the manuscript from many others: Malcolm Keating, Mohan Matthen, Adam Pautz, Susanna Schellenberg, Daniel Stoljar, and audiences at the 2020 and 2021 Online Colloquia in the Philosophy of Perception, the Australian National University, the Institute of Philosophy in London, Monash University, the National University of Singapore, the University of Cambridge, the University of Melbourne, the University of New South Wales, the University of Sydney, the University of York, and Yale-NUS College. I also thank my undergraduate students who worked through some very early drafts in my 2019 *Perception* seminar: Farheen Asim, Fred Chang, Siddharth Chatterjee, Zhi Heng Chia, Joseph Han (who also proofread an earlier version of the entire manuscript), Dara Hanson, Elaine Li, Allison Love, Yi Ming Ng, Gena Soh, Joshua

Tan, Yong Li Tng, Xianda Wen, Cassandra Woo, and Keith Yee. I must also acknowledge the detailed, and very useful, written feedback that I received from several anonymous referees.

I am deeply grateful to all of these readers. They have made the book much better than it otherwise would have been.

1 Rich Pluralism Introduced

You consciously see a ripe mango, one that is ovoid and orange, under mundane circumstances. The central question in philosophy of perception is: what is it for you to enjoy a conscious perception such as this?

1.1 An Unargued Assumption

I will shortly introduce a major assumption about how to answer this question. First, however, I must clarify a crucial notion: the notion of *awareness*.

Awareness can be characterized in relation to *knowledge*, which has four relevant features. First, knowledge is a relation between subjects and propositions. Second, instances of knowledge are, in and of themselves, free from error. That, together with the first feature, entails that one can only know true propositions, never false ones. Third, to know a proposition, a subject must have a special kind of *nonaccidental* (non-Gettiered) *connection* to it; accidentally true beliefs are not knowledge. Finally, knowledge is—I assume—a natural kind. In other words, the category of knowledge is metaphysically real, or structural, or joint-carving; it is not metaphysically arbitrary or gerrymandered.

I understand *awareness* as being broader than knowledge with respect to the first feature, while sharing the remaining features. In particular:

First, awareness can, in principle, relate subjects to entities of many different kinds: not just propositions but also objects, properties, states of affairs, and so on. Second, instances of awareness are, in and of themselves, free from error. That, together with the first feature, entails that awareness cannot be directed at false propositions, nonexistent objects, nonexistent properties, nonobtaining states of affairs, and so forth. (Note well: awareness *can* be directed at uninstantiated universals, as long as the subject does not *thereby* erroneously take the uninstantiated universal to be instantiated.) Third, to be aware of something, a subject must have a special kind of nonaccidental connection to it. Finally, awareness and its subkinds are natural kinds; the category of awareness is metaphysically real, or structural, or joint-carving, and so are any subkinds of awareness, such as knowing, remembering, and (I will suggest) perceiving.

Knowledge has a fifth feature: all instances of knowledge have *contents*. I will clarify the notion of content in chapter 5, but for now, note that I do *not* assume that all instances of awareness share this feature. Certainly *some* instances of awareness have contents: again, all instances of knowledge are, by definition, instances of awareness. But perhaps some instances of awareness—such as instances of perceptual awareness—do not have contents.

Now that we understand the notion of awareness, I can introduce an assumption that I will make throughout this book. It is this:

> The **framing assumption**: When you (consciously) see the mango, you are sensorily aware of it and its color and shape. In addition, whether you (consciously) see or hallucinate a mango, you are not sensorily aware of any peculiar mental entities.

(Throughout the book, bold type is used to mark terms that can be found in the glossary.) The framing assumption is supported by familiar arguments, and among contemporary philosophers of perception, it is widely, though not universally, accepted. Thus, I will provide scant defense of it. My focus will be on clarifying, in four parts, what the assumption means.

The first part of the assumption is that when you see the mango, you enjoy some form of *awareness*. This is sometimes known as the *act–object view*, and I find it plausible just on the basis of introspection. When I reflect on a conscious perception of an ovoid orange mango, it seems obvious to me that I am aware of *something*—a particular object and some of its properties.

This assumption has consequences for the *adverbialist* theory of perception. According to this theory, all that happens when you consciously see the mango is that you undergo certain mental modifications that are appropriately caused by the mango.[1] If these mental modifications do not involve awareness of anything, then the act–object view rules out adverbialism.[2]

The second part of the framing assumption is that when you see the mango, your awareness is *sensory* (rather than, say, cognitive). To appreciate the significance of this addition, consider the *dual-component* theory of perception, which says that when you consciously see the mango, you have

(i) a *sensation* that directly causes (ii) a *belief* that there is an ovoid orange mango in front of you.[3] The sensation is sensory, but it is not a form of awareness, and the belief might be a form of awareness—it might amount to knowledge—but it is not sensory. Thus, the dual-component theory is inconsistent with the second part of the framing assumption.

To unpack the rest of the framing assumption, we must distinguish between *direct* and *indirect* awareness. The basic idea is that to be *indirectly* aware of something is to be aware of it *by* being aware of something else; to be *directly* aware of something is to be aware of it, but not by being aware of something else. For instance, you might see a shadow and thereby become aware that there is a person rounding the corner. Plausibly, you are directly aware of the shadow but only indirectly aware of the person. There are many different ways of making this distinction precise, but for present purposes, we need not choose among them.[4]

I will simply use the term *sensory awareness* to refer only to *direct* awareness. Thus, the third part of the framing assumption is that when you see the mango, your awareness of it, and its color and shape, is *direct*. This view, too, is plausible on the basis of introspection. I do not seem to be aware of the mango, or its color and shape, *by* being aware of other things. I seem to be aware of them directly.

The fourth and last part of the framing assumption is that when you see a mango, or even just hallucinate one, you are *not* (directly) sensorily aware of peculiar mental entities. Perhaps your conscious perception of a mango can make you sensorily aware of the perception itself or of properties like the blurriness or sharpness of your vision: your perception and the properties of blurriness and sharpness are ordinary mental events and properties, not peculiar ones. But I assume that neither your perception of a mango, nor a hallucination of one, will make you sensorily aware of a private mental object that is ovoid and yellow. I also assume that the perception and the hallucination will not make you sensorily aware of internal "mental paints" instead of, or in addition to, external shapes and colors.[5] This assumption is supported by introspective and theoretical considerations. On the introspective side, seeing or hallucinating a mango does not *seem* to make you aware of peculiar mental entities.[6] On the theoretical side are well-known problems that I will not review here.

The third and fourth parts of the framing assumption are both violated by the once-popular *sense-datum theory*. This theory says that when you consciously see an ovoid orange mango, you are not sensorily aware of the mango itself, contrary to the third part of the assumption. Instead, you are sensorily aware of a peculiar mental entity, a *sense-datum*, that is ovoid and orange, contrary to the fourth part of the assumption. Thus, I set the sense-datum theory aside.

In sum, I assume without argument that when you see the mango, you are sensorily (and thus, by definition, directly) aware of it and its color and shape, but you are not sensorily aware of any peculiar mental entities. Given this framing assumption, what are our theoretical options?

1.2 Some Familiar Options: Representationalism and Naïve Realism

The contemporary philosophy of perception is dominated by two approaches: **representationalism** and **naïve realism**. For now, I will be loose in characterizing these theories, as well as the underlying notion of a representation. Increasingly precise definitions will be developed over the course of this book.

Representationalists think that consciously perceiving is purely a matter of having *representational* awareness of things, where *representations* are mental states or events that are subject to a certain kind of error. On this view, when you see the mango, you *successfully* represent it, in a distinctively sensory way, as being orange and ovoid. By contrast, when you hallucinate a mango, you deploy *failed* sensory representations of something that is orange and ovoid. Thus, there is a deep similarity between conscious perceptions and conscious hallucinations.

Naïve realists, by contrast, say that consciously perceiving is purely a matter of having *nonrepresentational* awareness of things—a kind of awareness that is immune to error, in a certain sense. The idea is that when you see the mango, you enjoy nonrepresentational awareness of it, its orangeness, and its ovoidness. There are many naïve realist accounts of hallucinations, but most of them agree that a matching hallucination of a mango does not involve nonrepresentational awareness of any object that is orange and ovoid. On these views, conscious hallucinations are very different in kind from conscious perceptions.

Representationalists and naïve realists disagree about whether consciously perceiving is a matter of having representational or nonrepresentational awareness, but buried under this disagreement is a shared assumption: on all versions of these views developed so far, it is agreed that there is *just one kind of sensory awareness* here. More precisely, both parties have—universally, as far as I know—accepted:

> **Monism:** What it is for you to consciously see, for example, an ovoid orange mango is for you to deploy *exactly one kind of sensory awareness*.[7]

Monism is really a cluster of views rather than a single one, for there is some vagueness in demarcating *kinds* of awareness. Still, we do not need a perfectly precise articulation of the monist thesis. Its spirit is clear: there is no major distinction in kind between, say, your awareness of the *mango* and your awareness of *ovoidness and orangeness*. With this understanding in mind, we can see what an enormously natural starting point monism is.

Widespread and natural as it is, I think that monism is false. I endorse:

> **Pluralism:** What it is for you to consciously see, for example, an ovoid orange mango is for you to deploy *two very different kinds of sensory awareness* in concert.

"Dualism" might be a more accurate label, but that term already has a well-entrenched meaning in the philosophy of mind. Thus, I have chosen a different label.

What drives me to pluralism is the idea—defended in great detail across this book—that when you see a mango, your awareness of the mango behaves *very differently* from your awareness of ovoidness and orangeness. Your awareness of ovoidness and orangeness reveals to you part of the very essences of those qualities, it generates hard problems of consciousness and the external world, and there are strong reasons to think that it occurs even in conscious hallucinations. By contrast, your awareness of the mango does not reveal anything about the essence of that object, it does not generate any hard problems of consciousness or the external world, and there is little reason to think that it occurs in conscious hallucinations. Pluralism lets us explain these asymmetries by positing one kind of awareness of the mango and a very different kind of awareness of ovoidness and orangeness.

It is possible to develop pluralistic versions of representationalism or naïve realism: we could concede that conscious perception consists of two very different kinds of sensory awareness, while maintaining that both kinds of awareness are representational or that neither are. However, I prefer a third option: I think that our awareness of particular mangoes is representational, while our awareness of ovoidness and orangeness is nonrepresentational. I call this theory **rich pluralism**.

In the next section, I give an overview of rich pluralism. The overview is dense and is not intended to be fully digested now. My aim is just to impart a holistic impression of the theory, so that you will know how its various pieces fit together as you advance through the book.

1.3 A New Option: Rich Pluralism

Rich pluralism is simply my very specific version of pluralism. I call it this because it is pluralism "enriched" with many further claims: if a thesis is defended anywhere in this book, then it is a commitment of rich pluralism.[8]

In this section, I will stake out three posits that are especially central to rich pluralism.

1. The rich pluralist begins by positing **deep awareness**, which has several distinctive features. First, deep awareness constitutes **hard consciousness**—roughly, the kind of consciousness that generates the hard problems of consciousness and the external world. It therefore occurs in all (hard-) conscious experiences, including conscious perceptions, hallucinations, sensory imaginings, and episodic memories. Second, at least in the sensory experiences of beings like us, the only targets of deep awareness are certain universals—the *sensory qualities*, which include colors and shapes. (What I am calling *targets* of awareness are more commonly referred to as *objects* of awareness. But I wish to avoid ambiguity: universals might be targets of awareness, but they are not objects, in the normal sense.)

   Third, deep awareness reveals a substantial part—not necessarily all!—of the essences of its targets. For example, when you consciously see a mango, you are deeply aware of ovoidness and orangeness. This reveals to you many essential truths about these sensory qualities: for example, that it lies in the nature of orange to be more similar to yellow than to green and to be instantiated only in something that is spatiotemporally extended.

   Fourth, deep awareness is nonrepresentational: it is, in a certain sense, immune to error. One consequence of this is that your deep awareness of orange cannot err about the essence of that color. Another consequence is that even when you consciously *hallucinate* something orange, you continue to have genuine deep awareness of orange, and a substantial portion of the very essence of that color continues to be revealed to you, just as in conscious perception.

2. The rich pluralist goes on to posit a second, very different kind of awareness: **successful sensory representation**. This contrasts with deep awareness in all of the respects just discussed. First, there is no constitutive link between successful sensory representation and consciousness. Successful sensory representation, *sans* deep awareness, equally occurs in the unconscious perceptions of simple creatures and super-blindseers. Second, whereas deep awareness only targets certain universals, successful sensory representation routinely targets particulars (and perhaps universals as well): when you see the mango, you successfully sensorily represent *it* and *its particular instances* of ovoidness and orangeness. Third, successful sensory representation does not reveal any part of the essences of its targets: seeing the mango does not reveal to you that it is essentially a fruit or even that it is essentially spatially extended.

   Fourth, successful sensory representation is, as the name suggests, a form of representation. It therefore has an error-ridden counterpart: part of what it is to hallucinate an ovoid orange mango is to deploy a *failed* sensory representation, one that merely *purports* to represent a particular mango and its particular instances of ovoidness and orangeness. Successful sensory representations of a mango and its property-instances constitute awareness of those targets; failed sensory representations do not.

   Although the rich pluralist thinks that deep awareness and successful sensory representation are very different, they also think that, in conscious perception, they are *woven together*. More precisely, the rich pluralist says that what it is to consciously see an ovoid orange mango is (i) to be deeply aware of universals such as ovoidness and orangeness; (ii) to successfully sensorily represent the particular mango and its particular instances of ovoidness and orangeness, via a perceptual relation to them; and (iii) to sensorily represent these property-instances *as* instances of the universals that the subject is deeply aware of. Condition (iii) is what ties together these different kinds of awareness; it is what makes any conscious perception a unified mental event.

3. The rich pluralist's theory has one last major piece, which describes something that is accidental to most experiences but essential to hallucinations. They say that on the basis of our deep awareness, our sensory

representations, and various contextual clues, we automatically and subpersonally *take* any experience to belong to certain kinds—for example, to be a perception, sensation, or sensory imagining. In simple creatures, this is a purely nonconceptual matter; it might consist of nothing more than simple behavioral dispositions. For sophisticated creatures like us, it also consists partly in *categorization*: the automatic, subpersonal application of various concepts, both to the targets of the experience and to the experience itself. In our focal case, you categorize what you see *as* an ovoid orange mango. More than this, you categorize the mango as a mind-independent object, and you categorize your experience as a perception.

These ideas about taking and categorization are hardly new. But the rich pluralist puts them to work in novel ways. They draw on them to explain why, in sophisticated subjects like us, perceptual experiences and sensations typically seem to *just present* particulars, which in turn seem to be *mind independent* or *mind dependent*, respectively. The rich pluralist says that these appearances are constituted by our categorizing our experiences in certain ways on the basis of their overall **gestalts**. They draw on these ideas again in their theory of hallucination: they say that what it is to be a hallucination is to be a sensory representation that is not a perception but is mistakenly taken for one.

That is rich pluralism in compressed form. The rest of the book will carefully unwind the theory.

1.4 The Structure of the Book

This book is divided into four parts.

Part I defends a thesis that I call **extended partial revelation**. This thesis is meant to characterize what is distinctive about *experience*, as opposed to, say, mere thought. The thesis states that, in the first instance, experience reveals to us a substantial portion of the very essences of certain universals, the *sensory qualities*, but it does not reveal any portion of the essences of anything else—for example, not of particulars.

I offer two separate arguments for extended partial revelation. Chapters 2 and 3 provide a direct argument: I gather many specific essential truths and show that extended partial revelation correctly predicts whether or not each one can be revealed in experience. Chapter 4 buttresses that with an indirect argument: I suggest that extended partial revelation best explains why some questions constitute hard problems of consciousness and the external world, while other, very similar questions do not.

Extended partial revelation is a crucial part of the motivation for rich pluralism. For this thesis suggests that when you see an ovoid orange mango, your experience makes you sensorily aware of *some* targets, like ovoidness and orangeness, in a way that is partly essence-revealing, while making you sensorily aware of *other* targets, like the mango, in a way that is not at all essence-revealing. There are many possible explanations of this difference. But, over the course of the book, I argue that the best explanation is that you are deploying two very different kinds of awareness in concert: a *partly* revelatory, *non*representational awareness of ovoidness and orangeness, together with a *non*revelatory, *representational* awareness of the mango and its instances of ovoidness and orangeness.

This book considers two main competitors to rich pluralism. (Most other competitors violate the framing assumption and are set aside.) The first competitor is *representationalism*, which I understand as the view that (i) to perceive is to deploy at least one kind of sensory awareness, typically of external particulars and their properties (or property-instances), and (ii) every kind of sensory awareness constitutive of perception is *representational* (i.e., subject to a certain kind of error). The second competitor is *naïve realism*, the view that accepts (i) but adds that (iii) every kind of sensory awareness constitutive of perception is *non*representational (i.e., immune to error, in a certain sense). As far as I know, all actual representationalists and naïve realists are monists. Still, in both of these definitions, clause (i) includes the phrase "at least one," so as to leave room for pluralistic versions of these theories.

Part II of the book—which consists of just a single chapter, chapter 5—argues that rich pluralism is superior to any form of representationalism, whether monistic or pluralistic. The basic thought is that the rich pluralist accepts all of the usual representationalist posits; they just add further posits about deep (i.e., partly essence-revealing) awareness. Thus, the rich pluralist gains automatic access to the core explanatory advantages of representationalism. In addition, rich pluralism is—surprisingly—just as parsimonious as adequate forms of representationalism. However, any form of representationalism faces a serious problem: because it says that our awareness of sensory qualities is representational, it predicts that experiences can err about the essences of the sensory qualities. In fact, this is impossible. Rich pluralists get things right: they say that our awareness of the sensory qualities is nonrepresentational and thus immune to this kind of error.

Part III of the book argues that rich pluralism is superior to any form of naïve realism, whether monistic or pluralistic. I start by identifying four major motivations for naïve realism. Naïve realism is thought to have an advantage in explaining why conscious perception (i) *reveals the intrinsic characters* of things (chapter 6), (ii) positions us to gain a certain special kind of *knowledge of reference* (chapter 7), (iii) simply *presents* the world to us (also chapter 7), and (iv) (correctly) *introspectively seems* to present the world to us (chapter 8). I argue that naïve realists have overlooked much of the data: many of the ideas above extend to conscious hallucinations, episodic memories, and sensory imaginings. I then develop the best accounts of the full range of data—where, lo and behold, several of these accounts require us to accept extended partial revelation. I go on to show that these accounts are compatible not only with certain versions of naïve realism but also with certain versions of representationalism and rich pluralism. I conclude that naïve realism brings no particular explanatory advantages to the table.

However, I show that if the naïve realist is to account for the full range of sensory events—including perceptions, hallucinations, sensory imaginings, and episodic memories, all of which come in conscious and unconscious varieties—then they must greatly complicate their theory. The rich pluralist's account is far more elegant, making it the better choice (chapter 9).

At this point, I conclude with some firmness that rich pluralism is superior to any other theory of perception.

Part IV of the book handles some remaining explanatory challenges—challenges that seem hard to meet on any theory of perception. In chapter 10, I address a deep problem about the *phenomenal character* of perception: it seems that this *must* be shared between conscious perceptions and matching hallucinations, and yet that it *cannot* be shared between them. I argue that philosophers have used the term "phenomenal character" with at least three different concepts and at least two different referents in mind. This dissolves the deep problem: there is one kind of phenomenal character that conscious perceptions and matching hallucinations must share and a different kind that they cannot share. In chapter 11, I rebut a series of arguments from hallucination. Chapter 12 concludes with a holistic summary of my case for rich pluralism.

Let me close with a methodological remark. Some philosophers seem content to argue that their theories of perception are supported by *some* compelling motivations and are not *refuted* by well-known objections. I am unimpressed by these arguments. The same can be said of *any* major theory of perception: motivations are abundant, refutations scarce.

This book makes more probing arguments. Regarding motivations, I show that *many converging ones* support rich pluralism and that the theory can honor *all* of the core motivations of its major competitors. Regarding objections, I show that rich pluralism can block *all* of the well-known ones *at little cost*. More generally, I try to develop the very best versions of my opponent's positions; to examine the full range of data, not just a prejudicial slice; and to openly admit the vices of my theory and the virtues of its rivals. I do all of this in search of the theory that is best, not just in some narrow respect, but overall.

Rich pluralism, I am convinced, is that theory.

I The Case for Partly Revelatory Awareness

There is something special about sensory experience. Seeing or vividly imagining scarlet makes you aware of *what scarlet is*, in some sense. Just thinking abstractly about scarlet does not do this. What is the difference?

In this part of the book, I answer that question with the following thesis:

> *Extended partial revelation*: In the first instance, all sensory experiences reveal a substantial portion of the essential truths about the sensory qualities and nothing else.

Extended partial revelation steers a path between two extremes. One is **full revelation**—roughly, the thesis that experience reveals the *full* essences of the sensory qualities. This thesis, I think, is too strong: experience is silent about *some* essential matters. This has driven many philosophers to the opposite extreme, **no revelation**, the thesis that experiences do not reveal *any* portion of the essences of anything. I believe that this thesis is too weak: it leaves us unable to explain why experience reveals *some*, indeed *many*, essential truths about the sensory qualities. Extended partial revelation is just the right thesis for the job.

I divide my defense of extended partial revelation across three chapters. In chapter 2, I directly defend a more limited thesis, **partial revelation**, which is the version of extended partial revelation that is restricted to conscious *perceptions*. In chapter 3, I extend this direct defense to *all* sensory experiences, including conscious hallucinations, sensations, sensory imaginings, and episodic memories. Chapter 4 buttresses this with an indirect defense of extended partial revelation. I observe that some questions constitute hard problems of consciousness and the external world, while other, very similar questions do not. I suggest that extended partial revelation best explains these asymmetries.

In chapter 3, I also consider which theories of perception can account for extended partial revelation. I argue that most monistic theories cannot do this, but some unusual monistic theories—**impure revelatory representationalism** and **impure revelatory naïve realism**—can; so can *pluralistic* versions of representationalism and naïve realism; and, of course, there is rich pluralism. Thus, a further conclusion of part I is that we should select from these theoretical options. Only in parts II and III do I argue for rich pluralism in particular.

2 A Defense of Partial Revelation

Perception makes us aware of visible qualities, such as scarlet; auditory qualities, such as middle C; tactile qualities, such as smoothness; olfactory qualities, such as the scent of vanilla; and gustatory qualities, such as the taste of peanuts. It also makes us aware of spatial qualities, such as sphericity, that are not tied to any particular modality. I will refer to such properties as the *sensory qualities*.[1]

Let us focus on cases of conscious perception. (This will be my exclusive focus in this chapter, so I will usually drop the word "conscious."[2]) In these cases, it seems that our perceptual awareness of sensory qualities is *substantive*, in some sense that is difficult to articulate.[3] For contrast, suppose that you just think abstractly of scarlet without experiencing anything scarlet. Although there might be some sense in which you are cognitively aware of scarlet, this awareness is not substantive in the present sense. It is clearly very different in kind from your awareness of scarlet when you actually *see* something scarlet. A central question of this chapter is the following: in what sense, precisely, does conscious perception provide *substantive* awareness of the sensory qualities?

I will build slowly toward my preferred answer, which takes the form of this thesis:

> *Partial revelation*: Consciously perceiving any sensory quality Q reveals a substantial portion of the essential truths about Q.

(Throughout the book, I will be unusually explicit: I will set off key claims from the main text—definitions, theories, arguments—much more than is usual. I believe that this promotes clear thinking, just as using Arabic numerals rather than Roman ones promotes quick arithmetic. To help the reader keep track of terms, there is also a detailed glossary at the end of the book.)

After explaining carefully what partial revelation means (section 2.1), I will suggest that this thesis is strongly supported by reflection on cases. I conclude that partial revelation tells us at least *part* of what is substantive about perceptual awareness (section 2.2). But is there anything more than this to the substantive character of perceptual awareness? If so, the answer must lie in a thesis that is even stronger than partial revelation. The obvious suggestion, and a very famous one, is *full revelation*—roughly, that perception reveals the *entire* essences of the sensory qualities (section 2.3). However, I argue in detail that any thesis in this neighborhood is vulnerable to serious objections—objections that do not touch partial revelation (sections 2.4–2.9).

This brings me to my conclusion and the thesis of this chapter (section 2.10). Partial revelation is not just true. It is also the best characterization of the sense in which perceptual awareness is substantive.

2.1 Partial Revelation Introduced

According to our focal thesis, partial revelation, consciously perceiving any sensory quality *Q* reveals a substantial portion of the essential truths about *Q*. The purpose of this section is to clarify what this means. Two notions here need explication: the notion of **essence** (and thus of **essential truths**) and the notion of *being revealed*.

On essences In a slogan, the essence of *x* is *what it is* to be *x*. This requires some unpacking.

We can begin with some plausible examples of essences. Mathematics teaches us that the essence of triangularity—what it is to be triangular—is to be a three-sided closed planar figure. According to utilitarians, what it is to be morally good is (roughly) to promote the greatest good for the greatest number. What it is to be gold, chemists say, is to be the element Au.

With these examples in mind, we can also identify some general features about essences. First, essences are had with metaphysical necessity: if what it is to be *x* is to be *F*, then, necessarily, all *x*s are *F*. Second, and more crucially, the reverse does not hold: not every metaphysical necessity pertaining to *x* is part of the essence of *x*. It is metaphysically necessary that Socrates is distinct from the Eiffel Tower, but this is no part of the essence of Socrates: what it is to be Socrates is not an architectural matter. Similarly, it is metaphysically necessary that Socrates is the sole member of the set

{Socrates}, and it is even part of the essence of {Socrates} to have Socrates as its sole member. However, it is *not* part of the essence of Socrates to be the sole member of {Socrates}. Socrates is not the individual that he is in virtue of being a member of a set! In short, essences are more fine-grained than metaphysical necessities.

Closely linked to the notion of essence is the notion of an essential truth: an *essential truth* about some entity x is simply a truth of the form <It lies in the essence of x that . . .>, where the angle brackets indicate contents. The relationship between (i) essential truths about x and (ii) truths about metaphysical necessities pertaining to x has been likened to the relationship between (iii) definitional truths featuring some term T and (iv) analytic truths featuring T. For one thing, definitional truths are more fine-grained than analytic truths. Not every analytic truth featuring term T is part of the definition of T: it is an analytic truth that triangles are three-sided closed planar figures, but this is no part of the definition of *planar*; it is rather the definition of *triangle*. This is parallel to the relationship that we have just noticed between metaphysically necessary truths and essential truths: it is metaphysically necessary that Socrates is a member of {Socrates}, but this is no part of the essence of Socrates; it is rather part of the essence of {Socrates}. One potential explanation of this parallel is that definitions are a special kind of essential truth—they are essential truths about the meanings of terms.

Perhaps the analogy between definitions and essential truths can be extended even further. There is *some* interesting relationship between analyticity and definitions. Plausibly, the analytic truths are simply the truths that obtain in virtue of the definitions of *all* terms. Similarly, perhaps there is some interesting relationship between metaphysical necessity and essence: perhaps the metaphysically necessary truths are simply the truths that obtain in virtue of the essences of *all* things.[4]

The skeptic might wonder why we should believe that there are such things as essences. One reason, I think, is that the notion does important work in metaphysics. But, so as to avoid large debates in other fields, I will not try to substantiate this point. A second reason is that the notion of essence is extremely intuitive. At least, I find it so, and in light of recent developments in metaphysics, it is clear that I have plenty of company.[5] This point will not persuade the skeptic, but it can still be a good reason for the nonskeptic to maintain their view. Finally, and most important, I will soon argue that perception reveals to us truths about the sensory qualities that are special in some way. In what way? I will argue that the best answer to this question appeals to essences. Other answers, such as those that merely appeal to metaphysical necessities, do not seem to do the job. This is another reason, and the most dialectically effective one, for accepting that there are essences.

On being revealed What is it for something to be **revealed**? My definition is this. For a mental state or event *M* to *reveal* a truth *p* about some entity *e* is for the following conditions to be met:

(i) *M* makes the subject aware of entity *e*.

(ii) For a conceptually sophisticated subject, rational reflection on *M* will, by itself, make it seem that *p* is true.

(iii) The subject's awareness of entity *e*, as described in (i), provides an intelligible basis for its seeming that *p* is true, as described in (ii).

There are several things to notice about this definition.

First, only mental states or events, and indeed ones that involve awareness of something, can reveal anything. Nothing can be revealed by nonmental entities, such as testimony, or by mental states or events that do not involve awareness of anything, such as false beliefs.

Second, the only things that can be revealed are *truths* (whether essential or **accidental**), understood as true propositions. Falsehoods cannot be revealed; at most, they can *mistakenly seem* to be revealed. Nor can nonpropositional items be revealed. Seeing a red apple does not reveal redness (or the apple); it reveals *truths* about redness. Still, for convenience, I will sometimes say that essences are revealed, when properly speaking what are revealed are truths about them.

Third, if a mental state or event reveals a truth *p* about some entity *e*, then *p* cannot seem to be true just brutely. The subject's awareness of *e* must provide an *intelligible basis* for *p*'s seeming to be true.

This last condition is the crucial one. To understand what it means, consider a case. I have never experienced any colors, and I see, for the first time, three different shades of red. Once I reflect on my experience, certain propositions, which are in fact true, will seem to me to be true—for instance, propositions about how these shades of red are intrinsically and how they relate to each other. Now imagine that you, too, have never experienced any colors, but one day, you have a *completely unconscious* perception of the same three shades, and that causes it to just *brutely seem to you* that the exact same propositions are true.

It is obvious that my epistemic position with respect to these propositions is much better than yours. What explains this? Not any difference in the propositions themselves: by stipulation, exactly the same propositions seem true to us both. However, one crucial difference is that for you, the propositions seem true *just brutely*: you have no idea *why* these particular propositions would be true rather than any other propositions. I, by contrast, having consciously experienced the three shades, can understand *why* these propositions are true. It is in this sense that my experience provides me with an *intelligible basis* for the seemings in question.

Fourth, the discussion so far might have given you the impression that whenever a mental state or event reveals a truth *p*, it must make *p* seem to be *immediately* or *obviously* true. Perhaps something like this is true of the ordinary word *revealed*, but in my technical sense, this is not the case. A truth *p* might be revealed even if *p* seems to be true only after complex feats of reasoning, imagination, and memory.[6]

Finally, to say that a mental state or event reveals a truth *p* is only to say that it makes *p* seem to be true *for a subject who is conceptually sophisticated* (among other things). What does this mean, precisely? Does the italicized expression refer narrowly to a subject who already possesses the concepts needed to entertain *p* or more broadly to one who is *capable* of forming those concepts? This turns out to be a delicate question. Let us return to it later in this chapter.

Now we have all of the pieces required to understand partial revelation. My next step is to argue that this thesis is true.

2.2 Partial Revelation Defended (and No Revelation Undermined)

The thesis to be defended is this:

> *Partial revelation*: Consciously perceiving any sensory quality Q reveals a substantial portion of the true claims of the form <It lies in the essence of Q that . . .>.

My argument for partial revelation begins with a case.

Here you are, seeing a scarlet pepper, an orange mango, and a green lime. You experience scarlet, mango orange, and lime green. Consider the following truths ("SQ" stands for "sensory quality"):

SQ1. It lies in the essences of scarlet and lime green that nothing can at once be wholly scarlet and wholly lime green.

SQ2. It lies in the essences of scarlet, mango orange, and lime green that scarlet is more similar to mango orange than it is to lime green.

SQ3. It lies in the essence of scarlet that any instance of scarlet is spatially extended.

I find it very plausible that SQ1–SQ3 will be revealed by nothing more than your perception. The most dramatic way to make the point is to imagine that you have never seen any colors before. Then you undergo a surgery that miraculously gives you ordinary sight, immediately letting you see like an ordinary person, without any adjustment period. It seems clear to me that simply seeing these three colored objects and considering your experience carefully will make SQ1–SQ3 seem true to you. That will be so even if you initially lacked the very concepts of scarlet, mango orange, and lime green: your perception will position you to acquire these concepts.[7] We could make similar points about cases in other modalities, such as cases in which you smell vanilla or hear middle C.[8]

That is my argument for partial revelation. An important rival to partial revelation is the following:

> *No revelation*: Consciously perceiving any sensory quality Q does not reveal any true claims of the form <It lies in the essence of Q that . . .>.[9]

No revelationists might raise several concerns about what I have said.

First concern My argument appeals to SQ2, the claim that it lies in the essences of scarlet, mango orange, and lime green that scarlet is more similar to mango orange than it is to lime green. Some will hold that SQ2 is false: science teaches us that color properties are, for example, surface reflectance properties, which do not stand in the similarity relations described by SQ2.

In response, I remain neutral about whether colors are surface reflectance properties, but I insist that they stand in the relevant similarity relations. After all, it seems perfectly plain that there are *metaphysically possible* properties, call them *color-like properties*, that are just the ways that colors seem to be: *possible* properties that have the same intrinsic characters and stand in the same similarity relations and that colors themselves perceptually seem to have and stand in. This much I cannot deny after reflecting on my perceptions. So either perception makes me aware of *actual instances* of these color-like properties, or it does not.

If perception makes me aware of actual instances of these color-like properties, then all is well: these color-like properties are surely just colors, and they have actual instances that stand in precisely the similarity relations that they seem to stand in. It might even happen that these actual instances of colors are surface reflectance properties, for the similarity relations among colors need not be the *physically salient* similarity relations that hold among these properties. They may instead be relations of *chromatic* similarity, which may be determined partly by the dispositions of perceivers such as ourselves. In that case, scarlet will still be more *chromatically* similar to mango orange than to lime green.

But suppose that perception does not make me aware of actual instances of these color-like properties; grant even that there are no such actual instances. Then these color-like properties are merely possibly instantiated. Still, it seems clear to me that I can think directly about them, just by attending to them as they show up in my perceptions. When I reflect on my perceptions of apparently colored things, I cannot find room for doubt on this point. Thus, my perception must make me sensorily aware of these color-like properties.

But how could perception make me sensorily aware of these color-like properties if they are never actually instantiated? This is possible if we say that I am sensorily aware of *uninstantiated universals* rather than property instances. And if perception makes me sensorily aware of color-like properties, understood as universals, and if these properties have the same intrinsic characters and stand in the same similarity relations as colors, then there is strong pressure to say that color-like properties *just are* colors.[10] So, once again, colors stand in the relevant similarity relations, and SQ2 remains true.

In sum, perception makes me sensorily aware of either *instances* of colors or *uninstantiated color universals*. Either way, SQ2 will end up being true: it will lie in the essences of scarlet, mango orange, and lime green that scarlet is more similar to mango orange than it is to lime green. That allays the concern.

Second concern According to SQ3, it lies in the essence of scarlet that any instance of scarlet is spatially extended. Some will worry that SQ3 cannot be true: this claim purports to be an essential truth about the *universal* scarlet, so it should not mention *instances* of scarlet.

I disagree. It is perfectly normal for an essential truth about one entity to mention other entities: the complete essential truth about the set {Socrates} will surely mention Socrates, the human being. Thus, it is reasonable, and indeed plausible, for essential truths about universals to mention particular instances. After all, these are instances *of those very universals*.

You might worry that on this view, the essence of a particular instance of scarlet will end up being the same as the essence of the universal itself. This worry is not well founded. It lies in the essence of an instance of scarlet to be spatiotemporally located. It does not lie in the essence of the universal scarlet to be spatiotemporally located but only that its instances must be spatiotemporally located.

Third concern Partial revelation says that perception reveals truths of the form <It lies in the essence of sensory quality Q that . . .>. You might wonder: why not just say that perception reveals to us truths about the sensory qualities that are *in fact* essential, without revealing *that* these truths are essential?

My answer is that after seeing the colors, we find ourselves very attracted to claims such as SQ1–SQ3, which explicitly refer to essences. It is mysterious what could have that effect on us besides reflection on perception.[11] Certainly, scientific inquiry does not do the job. If anything, the science of color makes it harder to see how the relevant claims could even be *truths*, much less *essential* truths. The only plausible option is that reflection on perception, all by itself, reveals truths of the form <It lies in the essence of sensory quality Q that . . .>.

Fourth concern I say that perception reveals many *essential* truths about sensory qualities. Why is it not enough to say that perception reveals many *metaphysically necessary* truths about sensory qualities?

I offer several arguments for the stronger formulation.

My first argument is a simple appeal to intuition. Look again at the scarlet peppers. Doesn't that seem to reveal to you at least part of *what it is* to be scarlet?

My second argument is similar to my response to the previous concern. Once we reflect on our perceptions, we find ourselves drawn to claims of the form <It lies in the essence of sensory quality Q that . . .>. Such claims receive little evidential support from claims of the form <Necessarily, . . . Q . . .>: it is necessary, but not essential to Socrates, that he belong to

{Socrates}. Thus, to say that perception only reveals necessary truths about sensory qualities is to leave it mysterious how we get to claims about their essences.

My third argument that the point should be expressed in terms of essences is that necessities are cheap. Take any case in which you know—not just through perception but via any method at all—that an entity *E* has property *F*. This already reveals to you infinitely many necessary truths about *E*: for example, that, necessarily, it is possible for *E* to exist; that, necessarily, it is possible for *E* to be *F*; and that, necessarily, *E* is such that $2+2=4$. These necessary truths seem, in some good sense, to be *cheap*. Perception reveals necessary truths that do not seem to be cheap. Not cheap in what sense? I answer: in the sense of being essential.

I offer one last argument that perception reveals many essential truths about sensory qualities. When you see something scarlet, what you seem to be aware of, in the first instance, is how scarlet is *in actuality*. This idea, difficult as it is to express precisely, is an intuitive one: you do not, in the first instance, seem to be aware of the modal profile of scarlet, of how scarlet is across all possible worlds. This generates a puzzle. For your perception reveals to you many truths about how scarlet is necessarily, where these necessary truths are not of the cheap variety described above. How, then, could your awareness of how scarlet is *in actuality* reveal to you how it is *in all possible worlds*? The puzzle is solved if seeing scarlet reveals essential truths about it. For, on the one hand, the essence of an entity is, plausibly, found in the actual world. Thus, there is no mystery about how your perception could connect you with it. And, on the other hand, an entity's essence has modal consequences, which means that learning essential truths about something can teach you how it is in all possible worlds.

Summary: A further result Now that these concerns have been met, I conclude that we should reject no revelation. We should instead accept partial revelation: if a subject (consciously) perceives a sensory quality Q, then a substantial portion of the essential truths about Q will be revealed to them. To unpack that more fully, if a subject (consciously) perceives a sensory quality Q, then (i) the perception makes the subject aware of Q; (ii) if the subject is rational and conceptually sophisticated, then reflection on their perception will, by itself, make it seem that these essential truths obtain; and, most important, (iii) the subject's awareness of Q, as described in (i), provides an intelligible basis for the appearance that these essential truths obtain, as described in (ii).

Given the truth of partial revelation, I would like to draw a further conclusion: that this thesis captures at least *part* of the sense in which conscious perception provides *substantive* awareness. I find this conclusion immediately plausible. If perception reveals the very essences of the sensory qualities to us, this certainly seems to be substantive in the sense at issue. But let us consider a pair of worries.

The first worry is that partial revelation cannot capture even part of what makes perception substantive *in general*. For partial revelation only identifies an unconditional feature of the perceptions *of conceptually sophisticated subjects*: only for them will the relevant essential truths seem to obtain (and, even then, only on rational reflection). However, surely *all* perceptual awareness is substantive, even for subjects who are irrational or conceptually unsophisticated. We need an unconditional feature to explain this.

In response, I grant that we need an unconditional feature to explain why all perceptual awareness is substantive. But I think partial revelation identifies such a feature: all perceptual awareness provides the subject with the *intelligible basis* for its seeming that certain essential truths hold. Only conceptually sophisticated subjects can *extract* the relevant seemings from this basis, and even then, complex rational reflection might be required. But the intelligible basis is there for every subject.

A second worry is that, at least on some views, our awareness of sensory qualities is not propositional. How, then, can it generate seemings about essential truths that *are* propositional?

My answer is that it can do so by making the subject aware of the intelligible basis for such seemings. I do not have a detailed account of how it is possible to make the transition from being aware of an intelligible basis to having it seem that certain propositions are true. But I find it plausible that there is some way to make this transition rationally, and I see no reason to doubt this.

Conclusion Partial revelation is not just true. It also captures *at least part* of what it is for (conscious) perception to provide awareness that is *substantive*. But is there any more to the substantiveness of perceptual awareness than this?

2.3 Full Revelation Introduced

If we want to say that there is, we will need to accept a thesis that is even stronger than partial revelation. The most obvious way to do this is to accept a thesis called **revelation**, although I will rename it *full revelation* for the sake of clarity. The basic idea of full revelation is that perception reveals the *entire* essence of any sensory quality, not just a substantial portion of it.

Full revelation is a very famous thesis: many theorists accept it, and I am unaware of anyone who rejects it while accepting partial revelation.[12] However, in the remainder of this chapter, I will argue that any claim in the neighborhood of full revelation is false. My conclusion will be that partial revelation is the *best* characterization of what it is for perceptual awareness to be substantive. The purpose of this section is to clarify what full revelation means and outline an argument for it.

To understand what full revelation means, consider two ideas. The first idea is that perception reveals the full essence of the sensory qualities. More precisely:

> *Self-intimation*: Consciously perceiving any sensory quality *Q* reveals every true claim of the form <It lies in the essence of Q that . . .>.

The second idea is that insofar as you are rational, perception will never lead you astray about the essences of the sensory qualities. More precisely:

> *Infallibility*: Consciously perceiving any sensory quality Q does not mistakenly seem to reveal any false claims of the form <It lies in the essence of Q that . . .>.

I stipulate that if a mental state or event *M mistakenly seems to reveal* some claim *p*, then *p* must seem to be true purely on the basis of *rational* reflection on *M*. It is irrelevant whether *p* seems to be true to a subject who is even slightly *irrational*.

Now we understand full revelation: it is simply the conjunction of self-intimation and infallibility.

The best argument that I can see for full revelation runs as follows. Suppose, as I have already argued, that partial revelation is true: seeing scarlet, for example, reveals a substantial portion of the essential truths about scarlet. Then it is initially plausible that it reveals all of them. For it is hard to see what essential truths about scarlet could be left hidden from us: there seems to be no truth about the essence of scarlet that is not laid bare by perception. So self-intimation is true.

Similarly, consider every claim about the essence of scarlet that seems to be true just on the basis of seeing scarlet and reflecting carefully, and in a fully rational way, on your experience. It is initially plausible that all of these essential claims really are true—that there are no essential falsehoods about scarlet that strike us as true at the end of careful reflection. Indeed, your experience seems to provide an *unimpeachable rational basis* for arriving at these essential truths. Thus, infallibility is also true.

Is this reasoning sound?

2.4 Partial versus Full Revelation: Gradually Acquired Concepts

In the next several sections, I will argue that self-intimation is false: (conscious) perception, powerful as it is, does not reveal *every* essential truth about the sensory qualities. I begin by reviewing several direct objections to self-intimation. Then I use these objections to pinpoint where our earlier argument for self-intimation went wrong.

Note that self-intimation will be my only target. None of my concerns will apply to infallibility. This will become important later: in chapter 5, I will defend infallibility and use it in an argument against representationalist theories of perception.

The goal of this section is to identify one cluster of compelling reasons to prefer partial revelation to its bolder sibling. My basic idea is that perception positions us to acquire new concepts *in a gradual way*, and the partial revelationist has a much easier time making sense of this than the full revelationist.

It is obvious that perception *sometimes* positions the subject to acquire new concepts: seeing scarlet for the first time will position the subject to acquire the concept of scarlet. The question, then, is, Which concepts, in general, does perception position the subject to acquire? A simpleminded suggestion is this:

> *Minimal concept acquisition*: Consciously perceiving sensory qualities $Q_1 \ldots Q_n$ positions the subject to acquire concepts of $Q_1 \ldots Q_n$, but it does not position them to acquire concepts of any other properties.

Minimal concept acquisition is plainly false. Suppose that you see a series of shades of red that are identical in hue and saturation but that gradually increase in brightness. This will position you to acquire not just concepts of those particular shades of red but also the concept of brightness.

Perhaps these experiences position you to acquire the concept of brightness because the shades of red that you see differ in brightness by their very essences. To put the proposal more generally:

> *Maximal concept acquisition*: Consciously perceiving sensory qualities $Q_1 \ldots Q_n$ positions the subject to acquire any concept needed to entertain any essential truth about $Q_1 \ldots Q_n$.

This is the most natural thesis to pair with full revelation. Unfortunately, it is false. Consider the limit case in which the subject sees something that is scarlet without seeing any other colors. It is essential to scarlet that it has a particular hue, saturation, and brightness. Still, this single experience will not, by itself, position the subject to acquire the concepts of hue, saturation, and brightness. The subject will be unable to differentiate these dimensions just by seeing a single color.

What seems to be true, then, is this:

> *Moderate concept acquisition*: Consciously perceiving sensory qualities $Q_1 \ldots Q_n$ positions the subject to acquire concepts of $Q_1 \ldots Q_n$ and *sometimes* positions the subject to acquire *some* of the concepts needed to entertain certain essential truths about $Q_1 \ldots Q_n$.

Accepting moderate concept acquisition lets the full revelationist explain what is happening when I see scarlet. Why doesn't my perception reveal to me that it is essential to scarlet that it has a particular hue, saturation, and brightness? Because my perception does not position me to acquire the concepts of hue, saturation, and brightness. Thus, I cannot even entertain the relevant thought about the essence of scarlet.

But there is a problem with combining moderate concept acquisition and full revelation: though these claims are logically consistent, it is hard to see how they could both be true. What on earth could be happening in a single perception so that it will reveal the entire essence of a sensory quality *if* the subject possesses the requisite concepts, and it will position the subject to acquire *some* of those concepts, but it will not position the subject to acquire *all* of those concepts?

What makes this question even harder to answer is that perceiving further sensory qualities *will* let the subject acquire all of the concepts in question. Seeing scarlet alone will not position the subject to acquire the concepts of hue, saturation, and brightness, but seeing many different colors will. It is very strange to say that individual perceptions are maximally powerful at revealing the essences of the sensory qualities, while for some reason being of intermediate strength in positioning the subject to acquire the concepts needed to think about those essences, even while a more complete sequence of perceptions is again maximally powerful at positioning the subject to acquire these concepts. Call this the *mismatch problem*.

The full revelationist might try to avoid the mismatch problem by saying that a subject must perceive at least two points along a single dimension (hue, saturation, or brightness) to form a concept of that dimension. This reply will not do. For consider sky blue and olive green. It is essential to these colors that they differ in hue, saturation, and brightness, so a single perception of these two colors will expose me to two different points along each of these three dimensions. But will this perception, all by itself, position me to form the concepts of hue, saturation, and brightness? Surely not. Much less will it reveal to me that it is essential to sky blue and olive green to differ across all three dimensions.

I do not know of a good way for the full revelationist to handle the mismatch problem. By contrast, the partial revelationist avoids the problem entirely: their view fits neatly with moderate concept acquisition. According to partial revelation, any perception reveals a substantial portion, but not necessarily all, of the essential truths about the sensory qualities experienced. Thus, the partial revelationist can say that any single perception, or set of perceptions, simultaneously reveals certain essential truths *and* positions the subject to acquire all of the concepts needed to entertain *those particular truths* (assuming that these concepts can be acquired through perception at all). On this view, there is no mismatch between what essential truths get revealed and what concepts the subject can acquire.

This is one important reason for preferring partial to full revelation.

2.5 Partial versus Full Revelation: Gradually Revealed Essences

Perception does not just position us to acquire new concepts in a gradual way. It also reveals essential truths to us in a gradual way. In this section, I argue that this observation, too, favors partial revelation.

Let us begin with a case. Suppose that I have seen enough warm colors (e.g., red, yellow, orange) to acquire the concepts of hue, saturation, and brightness. I then have my first perception of cool colors (e.g., blue, green, purple): an experience of sky blue and olive green. It is essential to sky blue and olive green that they differ in all three dimensions of the color quality space: hue, saturation, and brightness. And I already possess the concepts of hue, saturation, and brightness. Still, it seems that my perception will not reveal to me that sky blue and olive green differ in these precise respects. (This is already a counterexample to full revelation, but I will press this point in the next section. I am focusing on a different problem now.) Suppose that I go on to see many, many other colors—the entire Munsell color solid, perhaps. Then my experiences *will* reveal that sky blue and olive green differ in hue, saturation, and brightness.

If we accept full revelation, then this pattern is deeply problematic. Why do I need experiences of *other* colors to appreciate certain essential relationships that hold *just between sky blue and olive green*? The problem is not that I lack the concepts to entertain the requisite essential truths. In the present example, I possess those concepts. What, then, is the explanation? I cannot think of a good answer to this question. Indeed, I do not know of any theorist who has offered a systematic explanation of this curious pattern.

A great advantage of partial revelation is that it gives us at least the *outline* of an explanation. The thesis, you will recall, is that if a subject perceives any sensory quality *Q*, then a substantial portion of the essential truths about *Q* will be revealed to them. The basic thought is that if we accept partial revelation, we can say that perceptions of different colors gradually fill in more and more details about what is essential to a particular sensory quality.

You might wonder whether this is really possible: can seeing *new* colors truly help to reveal essential truths about colors that the subject has *already* seen? I think the answer is clearly yes. For consider these claims:

(i) It lies in the essence of Hypatia that she is a human being.

(ii) It lies in the essence of the property of a human being that any human being is a mammal.

Learning that (i) is true will not, by itself, tell us whether it lies in the essence of Hypatia that she is a mammal. We can arrive at this conclusion only if we also learn claim (ii). (We will also need an a priori "unpacking principle" about essences—namely, that if it lies in the essence of *a* that *a* is *F*, and it lies in the essence of *F* that any *F* is *G*, then it lies in the essence of *a* that *a* is *G*.) Claim (ii) is required even though it says nothing whatsoever about Hypatia. Thus it is possible, in principle, for essential truths that make *no* reference to a particular entity to help entail essential truths that *are* about that entity.

Conclusion Seeing new colors can position us to acquire, for the first time, concepts that apply to colors that we have already seen. It can also reveal to us essential truths about colors that we have already seen. If we accept full revelation, then it is difficult to see what could be happening in each individual color experience to make all of this possible. The partial revelationist can do better. They can say that seeing any particular color reveals only *some* essential truths about it, letting us deduce further essential truths upon seeing new colors. They can add that a single perception, or a series of them, positions us to acquire all of the concepts required to entertain whatever essential truths are revealed, if those concepts can be acquired through perception at all. This explains the patterns that we find.

Still, this leaves the partial revelationist with an important question: precisely what is revealed upon seeing a single color? I leave this as a question for future investigation.

2.6 Partial versus Full Revelation: Further Arguments

In this section, I offer further lines of support for partial over full revelation. The latter thesis comes in several subtly different versions. I will give a series of arguments that undermine each version of full revelation, while leaving partial revelation untouched.

First argument: Against simple full revelation Sensory qualities stand in certain relations of similarity: scarlet is more similar to mango orange than it is to lime green. Sensory qualities also stand in certain relations of quantity: this sound is louder than that one. The obtaining of these relations is evidently grounded in certain facts about the essences of the sensory qualities themselves. There are two influential models for how this might occur. The idea behind the first argument is that neither of these models is compatible with full revelation.[13]

Start with the *partial identity model*, which says that it is essential to the sensory qualities that they have certain internal structures. The idea is that scarlet is essentially composed of further subqualities (e.g., a certain hue, saturation, and brightness); it essentially has these subqualities as proper parts. Relations of similarity and quantity obtain between scarlet and other sensory qualities because these proper parts sometimes overlap.

Scarlet is more similar to mango orange than it is to lime green because scarlet and mango orange share more subqualities than do scarlet and lime green; one sound is louder than another because the first sound instantiates every determinate of loudness that the second does, as well as some further determinates.

The problem is this. Take a particular sound that is especially loud. It does not seem to be in the essence of that sound to instantiate a whole range of determinates of loudness. Thus, we have a counterexample to full revelation.[14]

There is another model for how to ground facts about similarity and quantity relations in facts about the essences of the sensory qualities. This is the *quality space model*. It says that any sensory quality is essentially a point in a multidimensional quality space. Scarlet, for instance, is essentially a point in a quality space whose dimensions include hue, brightness, and saturation. Any such quality space is holistic: no dimension of the space is metaphysically prior to the others, nor can any dimension of the space exist independently of the others. Degrees of similarity can then be analyzed in terms of distances between points in the same quality space, and quantities can be analyzed in terms of distances between points within a single dimension.

On this model, it is an essential truth about scarlet that it has a particular hue, brightness, and saturation, where these are in turn essentially located with the holistic color quality space. Suppose, then, that self-intimation is true. Then, once the subject sees scarlet, *even without ever having seen any other colors*, they should have access *to the entire color quality space*. They should be positioned to determine the precise nature of the dimensions of this space and thus all possible locations within it. That, in turn, should position them to recognize how scarlet is related to all other colors and indeed to identify all other colors. Evidently, this is not the case.[15]

Regardless of which model is true, then, the full revelationist is in trouble. There is, by contrast, no problem whatsoever for the partial revelationist. She does not say that seeing all of the colors will reveal to us their entire essences. Thus, it is no counterexample to their view if the grounds of similarity and quantity remain hidden from us, even after we have seen every color.

It is obvious that this argument undermines the simplest version of full revelation, namely:

> **Simple full revelation:** Consciously perceiving any sensory quality reveals all essential truths about it and does not mistakenly seem to reveal any essential falsehoods about it.

Still, some full revelationists have tried to dodge the argument by distinguishing *individual* from *collective* essential truths. Consider the essential truth that Socrates had such-and-such parents. This is an individual essential truth—an essential truth about Socrates, in and of himself. By contrast, consider the essential truth that Socrates is distinct from the Eiffel Tower. This is not an essential truth about either Socrates or the Eiffel Tower taken individually but about the collection of the two.

With this distinction made clear, the full revelationist might propose:

> **Individual full revelation:** Consciously perceiving any sensory quality reveals all *individual* essential truths about it and does not mistakenly seem to reveal any *individual* essential falsehoods about it.[16]

If we accept individual full revelation, then it will come as no surprise that experiencing scarlet does not, by itself, reveal how scarlet is related to all other colors. Such facts are not part of the individual essence of scarlet; they will be revealed only once the subject has seen the rest of the colors. Indeed, it is natural, though not mandatory, for the individual full revelationist to endorse:

> **Collective full revelation:** Consciously perceiving any number of sensory qualities reveals all *collective* essential truths about them and does not mistakenly seem to reveal any *collective* essential falsehoods about them.[17]

Collective full revelation has a nice formal feature and a nice substantive feature. The nice formal feature of collective full revelation is that individual full revelation drops out of it as a special case. The nice substantive feature is that collective full revelation explains why seeing many colors *would* reveal to the subject various similarity relations between scarlet and lime green.

To sum up: the argument from the grounds of similarity and quantity undermines simple full revelation. However, my opponent might avoid the concern by accepting individual and, if they wish, collective, full revelation.

Individual full revelation is the crucial thesis here, since its falsity would entail the falsity of collective revelation. Let us test it further.

Second argument: Against individual full revelation Individual full revelation entails that perception reveals the entire *individual* essences of the sensory qualities. Now, it is part of the *individual* essence of sky blue to have a particular hue, saturation, and brightness: that is what makes sky blue the individual color that it is. Thus:

AIR1. If individual full revelation is true, then seeing sky blue reveals to me that it is part of the individual essence of sky blue to have a certain hue, saturation, and brightness.

("AIR" stands for "against individual [full] revelation.") But does seeing sky blue—all by itself, for someone who has never seen any other colors—actually reveal this? Surely not:

AIR2. Seeing sky blue does not reveal to me that it is part of the individual essence of sky blue to have a certain hue, saturation, and brightness.

Thus:

AIR3. Individual full revelation is not true.

Is this argument sound?

I cannot bring myself to seriously doubt AIR2. It seems perfectly clear to me that if I see only sky blue, never having experienced any other colors, then I will not even be able to form the *concepts* of hue, saturation, and brightness on the basis of my experience. To confirm the point, feel free to inspect the patch of sky blue in figure 2.1 below:

Figure 2.1, Figure 2.2, and Figure 2.3
See also at http://profneilmehta.com/figures.html

The crucial question, then, is whether AIR1 is true.

It depends on how we interpret individual full revelation! This thesis requires the subject to be someone who wields concepts. Which concepts, though? The individual full revelationist might suggest that an essential truth *E* must be revealed only if the subject already possesses the concepts required to entertain *E*. Better yet, my opponent should stipulate that these concepts must have been acquired through perception, insofar as possible. In other words, they could say that what is true is not premise 1, but:

> AIR1*. If individual full revelation is true, then *if* I already have the concepts needed to entertain some individual essential truth about a sensory quality, where those concepts are derived from perception insofar as that is possible, then perceiving that sensory quality reveals that individual essential truth to me.

I observe that this does not save individual full revelation. For suppose that I have never seen any colors. I then have my first color perception, an experience of shades that are identical in brightness and saturation, differing only in hue: see figure 2.2 using the QR code above.

From this, I acquire the concept of hue. I then experience, for the first time, pure white: see figure 2.3 using the QR code above.

It turns out that white is *achromatic*: it has no hue. Will this be revealed to me, just on the basis of careful reflection on my limited color experiences? I submit that the answer is no. If all I have to go on are the limited color experiences just described, I cannot see how I could arrive at this conclusion. But we have built the example in such a way that I have a concept of hue, and it is part of the individual essence of whiteness that it has no hue. Thus:

> AIR2*. I already have the concepts needed to entertain some individual essential truth about a sensory quality (white), where those concepts are derived from perception insofar as that is possible, and perceiving that sensory quality does not reveal that individual essential truth to me.

Once more, the argument against individual full revelation goes through.

No such argument can be run against partial revelation, for there is no true analogue of the first premise. The entire point of partial revelation is that perception *needn't* reveal every essential truth about a sensory quality; it need only reveal *a substantial portion* of them. This is a powerful reason to prefer partial revelation to individual full revelation—or, for that matter, to collective full revelation, which is an even stronger thesis.

Conclusion In section 2.2, I gave an argument for partial revelation and asked whether we should go all the way to full revelation. It is now clear that we should not: we have found compelling arguments against every version of full revelation, whether simple, individual, or collective.

2.7 Holistic Revelation

The purpose of this section is to examine one more thesis in the neighborhood of full revelation—namely:

> **Holistic revelation:** Consciously perceiving *all* of the sensory qualities within a given quality space reveals all essential truths about those sensory qualities and does not mistakenly seem to reveal any essential falsehoods about them.

There is an important contrast between holistic revelation and the other theses about revelation that we have been considering. Return to the subject who sees their first color: scarlet. There is already something substantive about the kind of awareness that they enjoy, *even if* they expire before seeing any other colors. The central question of this chapter is, In what sense is perception substantive? Partial revelation says that it is substantive in the sense of partially revealing essences; individual full revelation says that it is substantive in the sense of fully revealing individual essences; collective full revelation, because it entails individual full revelation, says at least as much. By contrast, *holistic revelation does not answer the question*: it only tells us what happens when a subject perceives *all* sensory qualities within a quality space. To be fair, holistic revelation entails that a perception of scarlet has the conditional feature of helping to reveal all essential truths about all colors *if* it is conjoined with perceptions of all other colors. But this is not the feature that makes seeing scarlet substantive: the latter feature is not conditional on whether the subject sees anything else.

For this reason, I will not count holistic revelation as an interpretation of full revelation. (That is why I have not called it *holistic full revelation*.) I find it convenient to stipulate that full revelation is a thesis that, at a minimum, characterizes what is special about perceptual awareness. Still, holistic revelation deserves our attention. For this thesis retains something of the spirit of full revelation: it retains the idea that perception, by itself, can reveal the complete essences of the scarlet, middle C, and vanilla scentedness. At the same time, holistic revelation manages to avoid all of the objections to full revelation that we have seen so far. Unlike the simple revelationist, the holistic revelationist does not predict that if the quality space model is true, then seeing scarlet, by itself, will reveal the essences of all other colors. Unlike the individual or collective revelationist, the holistic revelationist does not predict that once I have an experiential concept of hue, seeing white will reveal to me that white is essentially achromatic. The holistic revelationist predicts only that seeing *all* colors will reveal to me that white is essentially achromatic. While I am not entirely sure that this prediction is correct, I will grant it for the sake of argument.

More generally, I have argued that partial revelation is true, but every version of full revelation is false. Holistic revelation is perfectly compatible with this conclusion. It could be that seeing an individual color reveals only part of its essence, while seeing all of the colors reveals their full essences. These observations lead to the central question of this section: is holistic revelation true?

I say that it is not. In support of this answer, I will review an argument that undermines not just holistic revelation but every version of full revelation that we have seen so far.

Observe that some colors are *compound*: every hue of these colors seems to be *composed*, in a special way, of other colors. Take orange, for example: every shade of it seems to be reddish and yellowish. By contrast, other colors—the canonical list being red, yellow, green, blue, black, and white—are *elemental*. That is, some shade of each of these colors does not seem to be composed of any other colors. Pure red, for instance, seems neither orangish nor purplish.[18] It is presumably part of the *individual* essence of any color whether it is elemental or compound.

This tells against holistic revelation, simple full revelation, individual full revelation, and collective full revelation. For any of these theses entails that seeing every color reveals at least the entire *individual* essence of each color. But this seems not to be the case: in the history of disputes about color, there was a recalcitrant, centuries-long disagreement—by theorists who were to all appearances attentive, well informed, and tolerably rational—about which colors were elemental and which were compound. Some theorists held that *all* colors were elemental; some held that green was a compound of yellow and blue; some held that brown was elemental (it is now usually thought to be a compound of yellow and black). What ended up mostly resolving the disagreement were *empirical* results about opponent color-processing mechanisms in the brain. The moral is that seeing all of the colors does not reveal whether each particular color is, as a matter of its individual essence, elemental or compound.[19]

Some will want to resist this objection. Some a priori truths are extremely difficult to discover: think of the centuries that mathematicians spent trying to get a handle on a priori truths about infinities. So perhaps seeing all colors *does* reveal which ones are essentially elemental or compound, and the historical parties just failed to notice this.

I reply that in the mathematical case, it is possible to demonstrate, conclusively, the inconsistencies or errors in various lines of reasoning. I cannot find anything like this in debates over the elemental–compound distinction. Precisely which inconsistencies or errors of reasoning were some parties to the debate supposed to be committing? Or which introspectively accessible truths were they failing to notice? Seeing no good answers to these questions, I believe that the objection is sound.

At least, it is sound when aimed at holistic revelation or any version of full revelation. The objection goes wide of *partial* revelation. This last thesis tells us only that seeing every color reveals *a substantial portion* of the essential truths about each color. It is not a problem for the partial revelationist if other essential truths, like those about which colors are elemental and which are compound, remain hidden.[20]

Summary We have examined the following readings of full revelation:

> *Simple full revelation*: Consciously perceiving any sensory quality reveals all essential truths about it and does not mistakenly seem to reveal any essential falsehoods about it.
>
> *Individual full revelation*: Consciously perceiving any sensory quality reveals all *individual* essential truths about it and does not mistakenly seem to reveal any *individual* essential falsehoods about it.
>
> *Collective full revelation*: Consciously perceiving any number of sensory qualities reveals all *collective* essential truths about them and does not mistakenly seem to reveal any *collective* essential falsehoods about them.

We have also considered the following thesis that, although not a version of full revelation as I am understanding things, retains something of its spirit:

> *Holistic revelation*: Consciously perceiving *all* of the sensory qualities within a given quality space reveals all essential truths about those sensory qualities and does not mistakenly seem to reveal any essential falsehoods about them.

I have argued that all of these claims are false. We should go as far as partial revelation, but no further. Now, I have already argued that partial revelation at least partly characterizes the sense in which conscious perception provides *substantive* awareness of sensory qualities. Given that no stronger thesis seems to be true, we can conclude that partial revelation is the *best* characterization of the substantivity of perceptual awareness of sensory qualities.

Some will worry that I have neglected some promising alternatives—in particular, (i) the view that perception makes us aware of the intrinsic characters of the sensory qualities, and (ii) the view that perception represents first-order and higher-order propositions about sensory qualities. I examine these views in the appendices to this chapter. In appendix 2A, I express qualified sympathy for the first view, but I suggest that even if it is true, it buttresses partial revelation rather than replacing it. As for the second view, I argue in appendix 2B that although it might seem initially promising, it does not survive reflection.

2.8 The Argument for Full Revelation Revisited

We have seen several direct arguments against full revelation. Still, we have yet to address the earlier argument that took us from partial revelation to full revelation. In this section, I pinpoint where that argument goes wrong.

The argument was this. When you look at a green lime and reflect, your perception will not merely seem to reveal *some* of what it is to be green. It will seem to reveal what it is to be green *in full*. There will seem to be nothing about the essence of green that remains hidden from you. Nor will your perception mistakenly seem to reveal any part of the essence of green. Thus, full revelation is required to do full justice to the introspective appearances.[21]

My response is to grant that things may introspectively seem this way *at first blush*. On reflection, however, I believe that this initial appearance is overthrown. We can come to appreciate this by reflecting on our color perceptions in light of some of the objections to full revelation. Here I am, looking at a green lime and experiencing green. It is obvious that green must be either essentially elemental or essentially compound. Which is it? I report that prolonged reflection on my experience does not even *seem* to deliver up an answer. On reflection, then, seeing green does not even *seem* to reveal its complete essence. Certain essential truths about it *do* seem to remain hidden from us.

On reflection, what the intuitions support is not full revelation but partial revelation.

2.9 The Retreat from Full Revelation

Full revelation decomposes into these two claims:

Self-intimation: Consciously perceiving sensory quality *Q* reveals every true claim of the form <It lies in the essence of *Q* that . . .>.

Infallibility: Consciously perceiving sensory quality *Q* does not mistakenly seem to reveal any false claims of the form <It lies in the essence of *Q* that . . .>.

Do the objections to full revelation undermine both of these claims or just one? In this section, I argue that the objections tell only against self-intimation. I offer two reasons for thinking that infallibility remains untouched.

First, I think there is a prima facie compelling case for infallibility. Perception does not just *brutely* make various essential claims seem true. It also provides us with an *intelligible basis*, something that helps us understand *why* these essential truths obtain in the first place. Seeing various colors teaches me not just *that* scarlet is essentially more similar to mango orange than to lime green but also *why this is so*. On reflection, I just cannot see how this intelligible basis could lead the subject astray *if* they are fully rational.

My second reason for accepting infallibility is that I cannot find any convincing counterexamples. A true counterexample would have to meet two conditions ("ICC" stands for "infallibility counterexample condition"):

ICC1. Consciously perceiving some sensory quality *Q* seems to reveal some claim *E* of the form <It is essential to *Q* that *p*>.

ICC2. *E* is false: it is not essential to *Q* that *p*.

Recall that for a perception to seem to reveal a claim, it is not enough for the claim to seem to be true to some subject just on the basis of reflecting on that perception. The subject must also be *fully rational*.

Bearing this in mind, I think that in all of the cases we have been discussing so far, we lack strong evidence that ICC1 obtains.

Notice that we do not, in general, get such evidence from cases of widespread *disagreement*—cases in which many think that perception *seems* to reveal *E*, while many others think that perception does *not* seem to reveal *E*. For in these cases, it is perfectly possible that perception *never* seems to reveal *E*, and the subjects who think otherwise are just overinterpreting their experiences. To get strong evidence that ICC1 obtains, we need a case in which there is widespread *agreement* that perception seems to reveal *E*.

This is absent in the cases we have been discussing so far. For instance, recall our discussion of the elemental–compound distinction. Some have argued that there is a counterexample to infallibility here: to *some* subjects, perception mistakenly seems to reveal that it lies in the essence of green to be compound, when in fact it lies in the essence of green to be elemental.[22] However, this is merely a case of widespread disagreement. Many subjects think that perceiving green seems to reveal that it lies in the essence of green to be elemental; others think that perception is silent on the question. Thus, those subjects who take green to be essentially compound might simply be irrational. The same is true of the other cases we have reviewed.

You might wonder whether a similar response could be used to defend self-intimation from my objections. The answer is no. For cases of widespread disagreement *do* support:

SICC1. Consciously perceiving some sensory quality *Q* does *not* seem to reveal some claim *E* of the form <It is essential to *Q* that *p*>.

SICC2. *E* is true: it is essential to *Q* that *p*.

("SICC" stands for "self-intimation counterexample condition.") If perception is ever *silent* about some essential matters, then both of these conditions are fulfilled at once. And we have very strong evidence that perception is silent about some essential matters: again, seeing sky blue and olive green perception does not, by itself, tell me how many dimensions of difference there are between these colors.

Thus, I reject self-intimation but accept infallibility. I will solidify my case for the latter thesis in chapter 5.

2.10 Conclusion

Most theorists of perception have chosen between the following theses:

> *No revelation*: Consciously perceiving any sensory quality *Q* does not reveal any true claims of the form <It lies in the essence of *Q* that . . .>.
>
> *Full revelation*: Consciously perceiving any sensory quality *Q* reveals every true claim of the form <It lies in the essence of *Q* that . . .> and does not mistakenly seem to reveal any false claims of the form <It lies in the essence of *Q* that . . .>.

I have argued that both of these theses are false. No revelation is too weak: perception reveals *many* essential truths. Full revelation is too strong: perception does not reveal *all* essential truths. What is true is:

> *Partial revelation:* Consciously perceiving any sensory quality Q reveals a substantial portion—not necessarily all!—of the true claims of the form <It lies in the essence of Q that . . .>.

Not only does partial revelation avoid the central class of counterexamples to full revelation, but it is also the true beneficiary of the intuitions that were supposed to favor full revelation. Even better, it provides the best account of the sense in which conscious perception provides *substantive* awareness of the sensory qualities.

Still, I accept the second half of full revelation, namely:

> *Infallibility*: Consciously perceiving any sensory quality Q does not mistakenly seem to reveal any false claims of the form <It lies in the essence of Q that . . .>.

We have found no compelling arguments against this claim so far, but my case for infallibility will not be complete until chapter 5.

Partial revelation is a thesis about the relationship between *perception* and the *sensory qualities*. In the next chapter, I extend this thesis: I argue that revelation occurs not only in perceptions but in all conscious sensory experiences—but, in the first instance, revelation *only* ever targets the sensory qualities; it never targets any particular objects, property instances, and so on. In the chapter after that, I offer a further line of support for this extended version of partial revelation: it is part of the best explanation of certain facts about the hard problems of consciousness and the external world.

Appendix 2A: Awareness of Intrinsic Characters

One central question of this chapter has been, In what sense does perception provide *substantive* awareness of the sensory qualities? My answer was: in the sense of being partly revelatory. In this appendix, I consider an apparently competing account that appeals to awareness of the *intrinsic characters* of the sensory qualities.

To motivate this apparent competitor, we can start with the observation that perceiving a sensory quality appears to entail some awareness of its intrinsic character. When you see scarlet, for example, you seem to be aware, at least in part, of what scarlet is like in and of itself, regardless of how it relates to anything else. This suggests:

> The *weak intrinsicality claim*: Consciously perceiving any sensory quality Q reveals a substantial portion of the intrinsic character of Q.

As we will see in chapter 6, I accept this claim, and in any case, it does not conflict with anything that I have said in this chapter. Indeed, the partial revelationist can even explain why the weak intrinsicality claim is true. They can say that these intrinsic aspects are essential to the sensory qualities, so, in revealing a substantial portion of the essences of the sensory qualities, perception can make the subject aware of these intrinsic aspects.

The trouble, however, is that once we accept the weak intrinsicality claim, we might be tempted to accept this bolder one, as well:

> The *strong intrinsicality claim*: The weak intrinsicality claim best characterizes the sense in which perceptual awareness of sensory qualities is substantive.[23]

This *does* seem to compete with my view that partial revelation best characterizes the sense in which perceptual awareness of sensory qualities is substantive.

We will need to trace a complex dialectic to get to the bottom of things.

The strong intrinsicality claim faces an objection: as I have argued, perception reveals a substantial portion of the essential truths about the sensory qualities, but the strong intrinsicality theorist has trouble explaining why this would be the case. For intrinsic truths need not be essential: my desk has its mass and shape intrinsically but accidentally. That very desk could have had a slightly different shape and mass.

Some strong intrinsicality theorists have developed an ingenious response to this objection.[24] Yes, they grant, not every intrinsic truth holds essentially. But it is a priori that intrinsic truths *within a special class* do hold essentially, and intrinsic truths about sensory qualities—truths of the form "It is intrinsic to sensory quality Q that *p*"—fall into this special class. These ideas deserve a meticulous examination.

The strong intrinsicality theorist begins with a pair of principles about *real* properties. We will worry later about exactly what a real property is:

I1. It is a priori that for any real property *G*, all possible instances of *G* resemble one another perfectly.[25]

I2. It is a priori that any two entities resemble one another perfectly just in case they are intrinsically identical.[26]

Assume that all logical consequences of a priori truths are themselves a priori. Then, I1 and I2 entail:

I3. It is a priori that for any real property *G*, all possible instances of *G* are intrinsically identical.

To express I3 in another way:

I4. It is a priori that for any higher-order property *H* of any first-order real property *G*, if *H* is an intrinsic property, then all possible instances of *G* have *H*.

We now arrive at the pivotal idea. The strong intrinsicality theorist advances a proposal about what essential properties are: they are simply *intrinsic properties that an entity has necessarily*. This proposal is moreover meant to be a priori:

I5. It is a priori that *F* is an essential property of *x* iff (i) *F* is an intrinsic property, and (ii) necessarily, if *x* exists, then *x* has *F*.[27]

At this point, the strong intrinsicality theorist has assembled all of the pieces needed to explain how perception positions us to know many essential truths about the sensory qualities. To illustrate the idea, suppose that purple is just a combination of bluishness and reddishness. Then, the strong intrinsicality theorist can say this:

I6. Seeing purple reveals to the subject that purple is a real property with the higher-order intrinsic properties of being bluish and being reddish.[28]

But I4 tells us that it is a priori that higher-order intrinsic properties of real properties are had necessarily, and I6 tells us that seeing purple reveals its higher-order intrinsic properties. So:

I7. Seeing purple reveals to the subject that (i) being bluish and being reddish are intrinsic properties, and (ii) necessarily, if purple exists, then it is bluish and reddish.

This satisfies the condition described in I5 for a priori knowledge of essentiality. Thus:

I8. Seeing purple reveals to the subject that being bluish and being reddish are essential properties of purple.

And truths like I8 are precisely what require explanation. In this way, the strong intrinsicality theorist hopes to do the requisite explanatory work. Call this the *intrinsicality-to-essence account*.

I find the intrinsicality-to-essence account to be elegant, even brilliant. How does it compare to partial revelation? Briefly, my answer is that the partial revelationist is welcome to accept this account, although it is not entirely clear whether they should.

An initial reservation is that the intrinsicality-to-essence account, as it was originally developed, was intended to entail full revelation.[29] However, I have already argued that full revelation is false. But there is an obvious way to weaken the account in light of this: the intrinsicality-to-essence theorist can retreat to the claim that perception reveals *a substantial portion* (not necessarily all!) of the intrinsic character of the sensory qualities. This is the version of the account that I have been considering throughout this section.

So understood, the intrinsicality-to-essence account is not in competition with partial revelation. Quite the opposite! The intrinsicality-to-essence account begins with claims about awareness of intrinsicality. It then uses these to derive claims such as I8, which are instances of partial revelation. Thus, the intrinsicality-to-essence theorist is offering an explanation of *why* partial revelation is true. The partial revelationist is welcome to help themselves to this explanation and to take the weak intrinsicality claim as a supporting piece of theory rather than a competing one.

I feel some temptation to do just that. But I am not quite sure that I should, for I see some objections. One objection is to I5—again:

I5. It is a priori that F is an essential property of x iff (i) F is an intrinsic property, and (ii) necessarily, if x exists, then x has F.

The left-to-right direction of this claim is subject to obvious counterexamples. It is often thought to be an essential property of Socrates that he had the particular parents that he did; it is often thought to be essential to any mammal that it have the right ancestry; it is often thought to be essential to my concept of water that it stand in a particular relation, perhaps historical or nomological, to actual water. None of these seemingly essential properties are intrinsic.

However, only the right-to-left direction of I5 is actually needed to derive partial revelation. I cannot immediately think of any counterexamples to this direction of the claim.

New problems arise when we turn to I6, which speaks of higher-order intrinsic properties of first-order properties such as purple. I find this talk somewhat puzzling. I understand what it is for a property *of a particular* to be intrinsic to it. I do not understand in general what it is for a higher-order property *of a property* to be intrinsic to it. Consider the property of *being a parent*. To have this property is to stand in the relation of parenthood to another. What is intrinsic to this property? I do not even know how to make sense of this question, much less answer it. Part of my puzzlement is that intrinsic properties are, roughly, properties that something has in and of itself, regardless of how it is related to other entities. But the property of being a parent *just is* a relational property—it requires being related to one's child—so I do not understand what it would mean to speak of what is intrinsic to this property.

But perhaps this problem can also be solved: perhaps it makes sense to speak of what is intrinsic or extrinsic *to an intrinsic property*. It is intrinsic to purple to be bluish but extrinsic to it to be the color of a famous fictional dinosaur. If this idea is adopted, and if purple is an intrinsic property—which it might reasonably be taken to be—then again the derivation might go through.

Conclusion I hold that perception reveals *both* part of the essences *and* part of the intrinsic characters of the sensory qualities. I see two promising theories of how these two kinds of revelation are related.

According to the first theory, what is fundamental is that perception partly reveals the essences of the sensory qualities. But, since it is essential to the sensory qualities to have certain intrinsic characters, perception also partly reveals these intrinsic characters. According to the second theory, what is fundamental is that perception partly reveals the intrinsic characters of the sensory qualities. Yet it is a priori that intrinsic properties have their intrinsic characters essentially. Thus, as a derivative matter, perception partly reveals the essences of the sensory qualities.

I remain neutral between these theories.

Appendix 2B: The Propositional Awareness Account

I have argued over the course of this chapter that perceptual awareness of sensory qualities is *substantive* in the sense of being partly revelatory. In this appendix, I consider an alternative account that analyzes the substantive character of perceptual awareness in terms of certain propositional attitudes. I argue that this account, though initially promising, is ultimately unsustainable.

The account is this:

> The *propositional account*: To consciously perceive something with a given sensory quality is to stand in a special relation, the relation of *sensorily entertaining*, to

> certain first-order and higher-order propositions involving that quality. This captures the sense in which perceptual awareness of sensory qualities is substantive.[30]

The idea is that to perceive something scarlet is to sensorily entertain various propositions, including not just the first-order proposition that the entity is scarlet but also the higher-order proposition that it lies in the essence of scarlet to be reddish and yellowish to certain degrees.[31]

The propositional account has several important advantages. It allows for an immediate explanation of partial revelation: the latter thesis is true because perceiving something with a sensory quality is a matter of sensorily entertaining certain propositions about that quality, *including some essential truths about it*. The propositional account also has the advantage of not entailing full revelation, a thesis that we have found to be too strong. Should we accept it, then?

I believe that the account faces a serious objection. It says that perceptual awareness of scarlet is substantive because the subject sensorily entertains various propositions about scarlet. For purposes of illustration, let us focus on the proposition that it lies in the essence of scarlet to be reddish and yellowish to certain degrees. I see a dilemma that turns on the following question: when we sensorily entertain this proposition, is our awareness of reddishness and yellowishness *itself* substantive or not?

On the first horn, the propositional theorist answers that this awareness is substantive. Since they say that awareness of a property *F* is substantive in virtue of the subject's sensorily entertaining propositions involving higher-order properties of *F*, they must now postulate that we sensorily entertain propositions involving third-order properties of these second-order properties. Is *this* awareness substantive or not? If so, then the propositional theorist must posit fourth-order awareness to explain that, and so on, ad infinitum. However, surely we do not have an infinitely layered awareness of this kind. Meanwhile, if the propositional theorist answers at any point that such awareness is *not* substantive, then they will be thrown onto the second horn of the dilemma.

On the second horn of the dilemma, the propositional theorist answers that no, our awareness of the relevant second-order properties is *not* substantive. The trouble is that substantive perceptual awareness cannot be assembled purely from pieces of awareness that are not substantive.

By analogy, suppose that I am sitting in a windowless room looking at a computer screen. Outside my room, there is a radar that scans the objects passing by and detects the degree to which each object has various gradable properties. I have no idea what properties these are. But as the radar does its work, all of this information is collected and sent to my computer screen. Now, the radar might tell me that it has just detected an object with property *F*, which essentially has 20 percent of property 1 and 83 percent of property 2. Next, it might tell me that it has detected a second object with property *G*, which essentially has 75 percent of property 1 and 54 percent of property 2. Finally, it might tell me that it has detected a third object with property *H*, which essentially has 97 percent of property 1 and 3 percent of property 2. Seeing the results of the radar will position me to know certain essential truths about properties *F*, *G*, and *H*, but my awareness of these properties will obviously fail to be *substantive* in the sense that we have been discussing. The problem is that this kind of awareness might make various essential truths seem true to me, but it will not provide me with the *intelligible basis* for understanding *why* they are true. The essential truths just seem to be true by brute force. That is not what our perceptual awareness of sensory qualities is like.

I conclude that the propositional account is false. Partial revelation better explains the sense in which our perceptual awareness of the sensory qualities is *substantive*.

3 Partial Revelation Extended and Explained

One central task of the previous chapter was to defend this claim:

> *Partial revelation*: Consciously perceiving any sensory quality Q reveals a substantial portion of the essential truths about Q, that is, truths of the form <It lies in the essence of Q that . . .>.

Partial revelation is a thesis about experiences of a certain kind, namely *perceptions*, and targets of a certain kind, namely *sensory qualities*. This makes it natural to wonder whether essential truths get revealed (i) by *experiences* of other kinds and (ii) for *targets* of other kinds. Regarding the first issue, I will argue that partial revelation extends to all sensory experiences, including sensory imaginings, episodic memories, hallucinations, and sensations (section 3.1). Regarding the second issue, I will argue that, in the first instance, essential truths get revealed only for the sensory qualities; secondarily, this also reveals some essential truths about instances of those qualities. However, experience does not reveal any essential truths about objects or high-level properties (like the property of being a pine tree) (section 3.2). In short, I will defend this thesis:

> *Extended partial revelation*: In the first instance, all sensory experiences reveal a substantial portion of the essential truths about the sensory qualities and nothing else.

This is a very specific thesis. What theories can explain it?

My preferred answer to this question comes in the form of a particular pluralistic theory that I call *rich pluralism*. After sketching some key parts of the theory (section 3.3), I show that it can shoulder the explanatory burden (section 3.4). I then argue that the rich pluralist's explanatory strategy can be co-opted by several other pluralistic theories and even by one unusual monistic theory, **impure revelatory monism**. However, all other monistic theories are incompatible with extended partial revelation (section 3.5).

At this point, I anticipate resistance: my explanation of extended partial revelation assumes that experience can make us sensorily aware of universals, but many objections have been fired at this assumption. I round out my discussion by offering my responses (section 3.6). I conclude that we should choose either (i) a pluralistic theory of perception, such as rich pluralism, pluralistic representationalism, or pluralistic naïve realism, or (ii) impure revelatory monism, which comes in representationalist and naïve realist varieties (section 3.7).

Of all these theories, it is rich pluralism that I prefer, for reasons that will be given in chapters 5–9. Still, the disjunctive conclusion of this chapter is a substantial one: the theories that remain in the contest are quite different from any previously developed theory of perception.

3.1 Experiences Other Than Perceptions

Let *sensory experiences* include perceptions, sensory imaginings, episodic memories, hallucinations, and sensations, as long as they are conscious. (Recall that *experiences* are, by definition, always conscious.) Partial revelation is only a thesis about conscious perceptions, but in this section, I will suggest that this thesis likely extends to *all* sensory experiences: they likely *all* reveal a substantial portion of the essences of sensory qualities. I think that the extension to sensory imaginings and episodic memories is clearly justified. For hallucinations and sensations, by contrast, I think that there are *pro tanto* reasons to make the extension, but we need further inquiry to reach a final verdict.

Sensory imaginings Suppose that you have never seen anything colored. Then neuroscientists perform a special surgery that enables you to *imagine*, in vivid detail, seeing colored things. After the surgery, you remain unable to see—your eyes are entirely nonfunctional—but you revel in your new abilities, imagining all sorts of colored things. At one point, you vividly imagine some patches of scarlet, lime green, and mango orange.[1] Does your imaginative experience reveal any essential truths about scarlet? As a reminder, what it is for a mental event *M* to *reveal* a truth *p* about some entity *e* is for the following conditions to be met:

(i) *M* makes the subject aware of entity *e*.

(ii) For a conceptually sophisticated subject, rational reflection on *M* will, by itself, make it seem that *p* is true.

(iii) The subject's awareness of entity *e*, as described in (i), provides an intelligible basis for its seeming that *p* is true, as described in (ii).

To my mind, it is obvious that your imaginative experience *does* reveal essential truths about colors. For example, upon rational reflection, it will seem to you that it lies in the essences of these colors that scarlet is more similar to mango orange than to lime green. What is more, your experience will provide you with an *intelligible basis* for these seemings.

Another way to appreciate the point is to imagine an ordinary subject who has seen plenty of colored things but has never reflected much on their experiences. After taking a metaphysics seminar, they acquire the concept of essences and are eager to think more about them. They close their eyes and begin to imagine different colored things: a scarlet dragon, a mango orange centaur, and a lime green orc. It seems clear to me that just on the basis of this imaginative experience, it will seem to them that it lies in the essences of these colors that scarlet is more similar to mango orange than to lime green. Again, this seeming will not be brute. The imaginative experience will provide an intelligible basis for it.

I offer one last observation to support the extension of partial revelation to sensory imaginings. There is an important difference between vividly imagining scarlet and actually seeing it: the sensory imagining will not make it seem that anything in your environment is scarlet; the perception will. Still, the sensory imagining seems to be *substantive* in the same sense as the perception. Just as there is an obvious difference between seeing scarlet and just thinking about it abstractly, there is the very same obvious difference between *vividly imagining* scarlet and just thinking about it abstractly. In the previous chapter, I argued that the substantiveness of the (conscious) perception is best characterized by partial revelation. Thus, if this same substantiveness is found in (conscious) sensory imaginings, we should extend partial revelation to cover them, too.

Episodic memories The extension of partial revelation to this case is straightforward—two of the arguments concerning sensory imagination apply just as well here—so I will be brief. First, we can consider an ordinary subject who has seen plenty of colored objects without ever thinking about their essences. Suppose that they acquire the concept of essence and then vividly remember many colored objects that they have seen. This alone will suffice to make it seem to them, on intelligible grounds, that scarlet is more similar to mango orange than to lime green. Second, episodic memory seems to be substantive in the same sense as perception and sensory imagination. I have already argued that this sense is best characterized by partial revelation.

Hallucinations Suppose that I merely hallucinate patches of scarlet, mango orange, and lime green. Should we accept the analog of partial revelation in such cases?

The intuitively plausible answer is yes. After all, my hallucination will introspectively seem to be exactly like a perception of these colored patches, and as I argued in the previous chapter, a perception of this kind meets (and introspectively seems to meet) conditions (i)–(iii) on revelation of essential truths. Thus, my hallucination will introspectively seem to meet conditions (i)–(iii), as well. This is some evidence that hallucinations really do reveal these essential truths.

There is a further observation that makes it especially tempting to say that hallucinations are partly revelatory. Consider an essential truth about colors that seems true, upon rational reflection, to the hallucinating subject—say, that it lies in the essences of these colors that scarlet is more similar to mango orange than to lime green. This claim will not strike the subject as just brutely true. Instead, they seem to be in a position to understand precisely *why* this claim is true: they seem to appreciate the *intelligible basis* that makes the claim true. This intelligible basis does not depend on whether anything in the subject's environment is actually scarlet, mango orange, or lime green; it just depends on those colors being *that way*, which they are. Thus, I find it very hard to see how this hallucination could fail to be partly revelatory.

Still, there seem to be theoretical pressures against the idea that hallucinations are partly revelatory. For example, when I *really see* a patch of scarlet, I am plausibly sensorily aware of a mind-independent patch of scarlet. When I *hallucinate* a patch of scarlet, it is hard to see how anything like this could be true: there is, we may suppose, no mind-independent patch of scarlet in my environment. In light of this, it is hard to see how my hallucination *could* reveal to me a substantial portion of the essence of scarlet. Relatedly, my hallucination is misleading in some respects: I seem to be sensorily aware of a mind-independent patch of scarlet, but I am not. So maybe I am also misled when I think that my hallucination partly reveals the essence of scarlet.

Notice that these same points do not apply to sensory imaginings and episodic memories. There are two crucial differences. First, there *is* a more obvious potential mechanism for how such experiences could be partly revelatory: this might occur because sensory imaginings and episodic memories are connected in some way to genuine perceptions, which we already know are partly revelatory. Second, sensory imaginings and episodic memories do not ordinarily mislead me into thinking that there are mind-independent sensory qualities around me when there are not. Thus, there is less reason to doubt the appearance that these experiences partly reveal the essences of sensory qualities.

By contrast, for hallucinations, the intuitive and theoretical considerations seem to pull in different directions. Thus, we cannot yet firmly say whether hallucinations genuinely reveal a substantial portion of the essences of sensory qualities. It is to the credit of a theory to say that hallucinations are partly revelatory in this way, but the overall best theory might end up saying that hallucinations merely *seem* to be partly revelatory. In other words, it is a desideratum, rather than a datum, to say that hallucinations are partly revelatory. I will argue in this chapter that rich pluralism (among other theories) gives us this desideratum. In later chapters, I will argue that rich pluralism also does justice to the very plausible claims that seem to be in tension with having this desideratum. This makes for an especially satisfying theory.

Sensations Most of what we have said about hallucinations can be carried over to sensations. Imagine experiencing a vivid scarlet afterimage. This certainly seems to reveal as much about the essence of scarlet as seeing a real scarlet thing. It is not just that the same claims about the essence of scarlet seem to be true; it is that you seem to have the same kind of intelligible basis for understanding *why* those claims are true. At the same time, there are theoretical pressures that push in the opposite direction. When you experience the afterimage, there is no mind-independent instance of scarlet for you to be aware of, so it is hard to see what mechanism could possibly reveal anything of the essence of that property. And the sensation arguably seems to make you aware of a mind-*dependent* instance of scarlet, but it is somewhat hard to believe that such instances really exist.

Once more, then, the intuitive pressures compete with the theoretical ones. Thus, it is *pro tanto* desirable to say that sensations are partly revelatory, but the best overall theory might end up rejecting this claim.

Conclusion Sensory imaginings and episodic memories reveal a substantial portion of the essences of their targets. It is desirable to say the same of hallucinations and sensations, though we will have to see whether this view is sustainable. Still, the most initially attractive thesis is that *all* sensory experiences reveal a substantial portion of the essences of sensory qualities.

3.2 Targets of Experience Other Than Sensory Qualities

We do not just experience sensory qualities. We also experience objects, perhaps *high-level properties*, such as the property of being a pine tree, and *instances* of sensory qualities (as opposed to universals). Is anything like partial revelation true of these other targets of experience? The purpose of this section is to defend a qualified negative answer to this question.

Essences of objects You are looking at a scarlet pepper. Discard all of your background knowledge and consider what seems true to you about the pepper purely on the basis of seeing it. Your perception reveals plenty of *accidental* (i.e., nonessential) truths about the pepper, such as its color and shape. But does it reveal any *essential* truths about it? I will argue that it does not.

Among contemporary metaphysicians, there are two especially prominent views about what is essential to objects. One is *sortal essentialism*, the view that an object could not have been of a radically different kind—at least for certain kinds—than it in fact is.[2] The sortal essentialist might say this about the pepper (letting "O" stand for "object"):

O1. It lies in the essence of the pepper that it is a fruit.

If O1 is an essential truth, it is not revealed by your perception alone. For all your perception tells you, the pepper might essentially be a gigantic atom, an infinitely divisible crystal, or a Martian in repose.

A second prominent view about the essences of objects is *origin essentialism*. This is the view that an object could not have had a radically different origin than it in fact has.[3] If the pepper was grown from seed *S*, the origin essentialist might assert:

O2. It lies in the essence of the pepper that it originates from something similar to seed *S*.

Again, however, your perception will be silent about the truth of O2. Looking at the pepper will not reveal to you that it came from a seed at all, much less that it did so essentially. In fact, consider a very minimal essential truth about the pepper, such as:

O3. It lies in the essence of the pepper that it is spatially extended.

Even this is not revealed by your perception of the pepper. It is compatible with your perception that *this very object* might be able to turn into a dimensionless point or a packet of energy with no particular location.

Still, you might wonder whether perception reveals more sophisticated essential truths about the pepper. For example:

O4. It lies in the essence of the pepper that it is *sometimes* spatially extended.

O5. It lies in the essence of the pepper that it is *possibly* spatially extended.

O6. It lies in the essence of the (red) pepper that it is *possibly* yellow.

O7. It lies in the essence of the pepper that it is self-identical.[4]

Aren't these essential truths revealed by perception? I think not. As we saw earlier, essential truths are truths about *what it is* to be something—what *makes* something the thing that it is. I think that this shows that O4–O7 are not even true and thus cannot be revealed.

One problem with O4 is that what is essential to that pepper—what it is to be that pepper—is not a matter of how it *sometimes* is (except perhaps at its origin). Similarly, one problem with O5 and O6 is that what it is to be this pepper, what makes the pepper the thing that it is, is a purely actual matter. It is not a matter of how it *possibly* is. I grant that these truths about temporality and possibility might be *necessitated* by essential truths. But they are not essential truths themselves.

There are, by the way, additional problems for O4 and O5. Regarding O4, even if this claim were true, the objection would fail. For your perception might reveal that the pepper is sometimes spatially extended, but it does not reveal *that this is essential to the pepper.* It is perfectly compatible with the deliverances of perception that this very object could *always* be a packet of dimensionless energy. Meanwhile, regarding O5, there is an additional reason to think that this claim is false: being colored seems to be entirely irrelevant to the essence of the pepper. What makes something a pepper is not its range of actual or possible colorings. (I do not think that there is an analogous problem for O6, since I find it plausible that being spatially extended is part of what makes something a pepper.)

Now return to:

O7. It lies in the essence of the pepper that it is self-identical.

I think that this, too, is false. It is much more plausible that it lies in the essence of *identity* that every entity is self-identical. The essence of the *pepper* seems to have nothing to do with it.

But suppose for the sake of argument that O7 were true and that perception revealed it to you. Still, this would not support the analogue of partial revelation for objects. For O7 is not a *substantial* essential truth: if the pepper is essentially self-identical, then so is everything! This is quite different from what happens with sensory qualities like scarlet and mango orange: what perception reveals about them is very substantial indeed. Thus, we could better explain why perception revealed O7 to you by saying that perception merely positions you to know that the pepper *exists*, and it is a priori that whatever exists is essentially self-identical. In support of this explanation, observe that *any* method of knowing that the pepper exists will position you to know that it is essential to the pepper to be self-identical.

I conclude that perception does not reveal any (substantial) essential truths about objects.

Essences of high-level properties and their instances Contrast seeing a pepper *for the first time* with seeing a pepper *after you have learned to recognize peppers*. These experiences are phenomenologically different. According to some theorists, the difference consists partly in the fact that after you have learned to recognize peppers, you experience not just low-level color and shape properties but also the property of being a pepper. More generally, these theorists hold that we can experience *high-level properties*, like the property of being a pine tree, a Russian sentence, or a smile.[5]

Suppose for the sake of argument that this view is correct. Should we then say that experience reveals a substantial portion of the essences of high-level properties or their instances? I think the answer is a clear *no*. Even after you can visually recognize peppers, seeing one seems to tell you nothing about what it is to be a pepper. For all you can learn from your experience, what it is to be a pepper is to be an alien spaceship or an insect cocoon.

Essences of instances of the sensory qualities When you see the pepper, your experience reveals part of the essence not only of the *universal* scarlet but also of *all particular instances* of scarlet. For example, it reveals that it lies in the essence of any *instance* of scarlet to be spatially extended and to be more similar to instances of mango orange than to instances of lime green. At the same time, seeing the pepper does not by itself reveal any *differences* in essence between the distinct instances of scarlet that you see on its surface. Your experience will not reveal anything about what it is to be this instance of scarlet *rather than that one*. (Perhaps the essential difference is a difference in haecceities, or in origins.) This is curious. It is different from what we find with the sensory qualities themselves: seeing scarlet *does* reveal to you part of what it is to be scarlet rather than lime green.

What is even more interesting is that it is initially plausible that hallucinating a scarlet pepper, or having a sensation of a scarlet afterimage, will reveal to you just the same essential truths about particular instances of scarlet. It seems that even your hallucination and your sensation will reveal to you that it lies in the essence of any instance of scarlet to be spatially extended and to be more similar to instances of mango orange than to instances of lime green. That is surprising: when you have a hallucination or a sensation, you are not encountering any mind-independent instances of scarlet. In the end, we may be forced to reject the view that such experiences are substantially revelatory about instances of sensory qualities, for reasons that we saw in our discussion of hallucinations and sensations in section 3.1. Still, this is the most initially attractive view.

In short, all sensory experiences seem to reveal some essential truths about instances of the sensory qualities—even hallucinations and sensations, which do not relate the subject to any mind-independent instances of these qualities. However, there is a curious limitation: nothing is revealed about what distinguishes different instances of the same sensory quality.

There is a natural way to explain these odd patterns. We can say that, in the first instance, sensory experience reveals essential truths only about sensory qualities, understood as *universals*. This reveals essential truths about *all particular instances* of those universals for free: it is a priori that all instances of a universal share certain aspects of the essence of the universal itself. For example, your experience of a scarlet pepper reveals to you that it lies in the essence of the universal scarlet that any instance of it is spatially extended. It is a priori that if this is true, then it lies in the essence of any instance of scarlet to be spatially extended.

This explains how hallucinations and sensations can reveal essential truths about instances of scarlet, despite not relating the subject to any *instances* of scarlet: they can still relate the subject to the *universal* scarlet. This also explains why experience never reveals any essential differences between distinct instances of scarlet: only essential truths about the universal scarlet are revealed. Though this will a priori entail some essential truths common to all instances of scarlet, it will not a priori entail anything about the essential differences between distinct instances of scarlet.

Summing up and looking forward Experience reveals a substantial portion of the essences of sensory qualities. As a consequence, it reveals certain limited truths about the essences of *instances* of sensory qualities—in particular, only those essential truths that fail to distinguish between different instances of the same sensory quality. Finally, when it comes to particular objects, high-level properties, and instances of high-level properties, experience is entirely silent about their essences. These ideas can be encapsulated in the following thesis:

> *Extended partial revelation*: In the first instance, all sensory experiences reveal a substantial portion of the essential truths about the sensory qualities and nothing else.

This is a very specific thesis. What theories can explain it?

In sections 3.3–3.4, I begin to sketch my preferred pluralistic theory of perception, and I show that it neatly explains extended partial revelation. In section 3.5, I argue that most monistic theories cannot do the same, but I identify several other pluralistic theories and one unusual monistic theory that can mimic my preferred explanation.

3.3 The Beginnings of Rich Pluralism

According to any pluralist, what it is to consciously perceive is to deploy (at least) two very different kinds of sensory awareness. In this section, I begin to lay out my preferred pluralistic theory, *rich pluralism*.

Rich pluralism begins with three posits about *deep awareness*:

DA1. Deep awareness occurs in all *sensory experiences*, including all conscious perceptions, hallucinations, sensations, sensory imaginings, genuine episodic memories, and merely apparent episodic memories.

DA2. At least in the sensory experiences of beings like us, the only targets of deep awareness are certain universals—the *sensory qualities*.

DA3. Deep awareness is a nonrepresentational form of awareness that reveals a substantial portion—not necessarily all!—of the essences of its targets.

To this, the rich pluralist adds three more posits about an entirely separate form of sensory awareness, *successful sensory representation*:

SSR1. Successful sensory representation occurs in all perceptions and in all (genuine) episodic memories, whether conscious or unconscious.[6] It does not occur in any hallucinations, sensory imaginings, or failed episodic "memories"—here we find only failed sensory representations.

SSR2. The targets of successful sensory representation include particulars, such as ordinary objects and their property-instances. (Rich pluralism is neutral on whether these targets sometimes include universals, too.)

SSR3. Successful sensory representation is a representational form of awareness that never even seems to reveal any portion of the essences of its targets.

Rich pluralism includes many further posits—the theory consists of every claim defended in this book—but these six posits form the theory's core.

Let me unpack the three posits about deep awareness, DA1–DA3. For now, I will let the three posits about successful sensory representation, SSR1–SSR3, serve primarily as a foil; they will not be fully unpacked until chapter 5.

DA1 tells us when deep awareness occurs: not just in perceptions but in all sensory experiences, where any *experience* is by definition conscious. I stipulate that *sensory* experiences include conscious perceptions, hallucinations, sensations, sensory imaginings, and episodic memories; sensory experiences can be contrasted with mere abstract thoughts. The idea is that I am deeply aware of scarlet—really aware of it, not just apparently so—whether I am consciously seeing, hallucinating, or imagining something scarlet. By contrast, SSR1 tells us that successful sensory representation occurs only in perceptions (including illusions) and genuine episodic memories. It does not occur in hallucinations, sensations, sensory imaginings, or failed episodic "memories."

The function of DA2 is to identify the targets of deep awareness: in the sensory experiences of beings like us, the targets are always universals. However, not every universal is a possible target of deep awareness. DA2 tells us that its only possible targets, at least in our sensory experiences, are the sensory qualities. Sensory experience never makes us deeply aware of high-level properties, such as the property of being a pepper.

This is a further difference between deep awareness and successful sensory representation: SSR2 tells us that the targets of successful sensory representation also include *particulars*, such as objects and property-instances. (Note that property-*instances* are particulars: they are spatiotemporally located, and they can be qualitatively identical but numerically distinct. Universals such as redness and roundness are not particulars.) Successful sensory representations will usually also represent those targets *as* similar and different in various respects. Perhaps this involves representing property-instances *as* instances of universals, or perhaps it just involves representing property-instances as being similar and different in certain ways.

DA3 characterizes the ontological structure and epistemic significance of deep awareness. Regarding ontological structure, DA3 begins by telling us that deep awareness is not propositional; it is awareness *of* (for instance) the universal scarlet, rather than awareness *that* scarlet is a particular way. SSR3 tells us that successful sensory representation contrasts with deep awareness in this respect. To this, SSR3 adds that successful sensory representation is, as the name suggests, *representational*. I will explain in chapter 5 precisely what I mean by this, but the basic idea is that any representation is *about* something. The thing that it is about is its *content*: a way for things to be. It is also a defining characteristic of representations that they can go wrong: you can stand in any given representational relation (e.g., belief) to a *false* or *truth-valueless* proposition, a nonexistent object, or a nonobtaining state of affairs. DA3 adds that deep awareness is nonrepresentational: it cannot go wrong in this way. Even in hallucination, the subject enjoys genuine deep awareness of universals.

In addition to all of this, DA3 says that deep awareness *reveals* a substantial portion of the essences of the sensory qualities. I have already explained in detail what this means.

Let us start to put these pieces together. Even though deep awareness and successful sensory representation are very different, the rich pluralist does not think that these kinds of awareness are utterly disjointed in perception. Rather, they say that when you see a scarlet pepper, you exercise these forms of awareness in concert. In particular, you successfully sensorily represent the pepper and its instances of scarlet, you are deeply aware of the universal scarlet, and you sensorily represent these property-instances *as* instances of the universal, scarlet, that you are deeply aware of.

How does rich pluralism compare to familiar versions of representationalism and naïve realism? The rich pluralist's posits about successful sensory representation will be entirely familiar to representationalists, especially to those who posit singular content.[7] Then there are the rich pluralist's posits about deep awareness. There is a certain scent of naïve realism here, insofar as the rich pluralist regards deep awareness as a *nonrepresentational relation* of awareness. But the rich pluralist also embraces an idea that many naïve realists are hostile to: the idea that the targets of deep awareness are universals rather than particulars.

Now we understand some of the major pieces of rich pluralism. Let us see the theory in action.

3.4 The Rich Pluralist's Explanation of the Asymmetries

In previous sections, we discovered some complex asymmetries, summed up by extended partial revelation, about which essential truths are revealed in experience, and we considered a novel theory of perception, rich pluralism. In this section, I argue that the theory neatly explains the asymmetries.

First asymmetry We have seen that perception reveals a substantial portion of the essential truths about sensory qualities. By contrast, it reveals no essential truths about objects.

Rich pluralism explains this asymmetry. DA1 tells us that we have deep awareness in cases of perception, and DA3 adds that deep awareness of any target *x* reveals a substantial portion of truths of the form <It is essential to *x* that ___>. DA2 then tells us that the only targets of deep awareness are the sensory qualities. Yes, in perception, we *successfully sensorily represent* ordinary objects and property-instances, as per SSR1 and SSR2, but SSR3 tells us that successful sensory representation is not even partly essence-revealing. That is why a veridical perception of a pepper, a mango, and a lime will reveal to you many essential truths about scarlet, mango orange, and lime green without revealing to you any essential truths about peppers, mangoes, or limes.

Second asymmetry Suppose that instead of really seeing these fruits, you merely hallucinate them. We have seen that it is *pro tanto* desirable to say that your experience will still reveal to you the same essential truths about scarlet, mango orange, and lime green. However, your hallucinatory experience, like the perception, will fail to reveal any essential truths about particular fruits.

The rich pluralist explains this asymmetry in the same way as before. DA1 tells us that we have deep awareness in all sensory experiences, including hallucination. DA2 says that this is awareness of sensory qualities, nothing else—in particular, not objects such as peppers. DA3 adds that this awareness is partly essence-revealing.

Third asymmetry Turn now to property-instances. We have found that seeing the pepper reveals to you part of the essence not only of the *universal* scarlet but also of *all particular instances* of scarlet. For example, it reveals to you that it lies in the essence of any instance of scarlet to be spatially extended. At the same time, seeing the pepper does not by itself position you to identify any *differences* in essence between the distinct instances of scarlet that you see on its surface. It will not reveal to you anything about what it is to be this instance of scarlet *rather than that one*. Finally, we saw that it was plausible, though not mandatory, to say that hallucinating a scarlet pepper or experiencing a sensation of a scarlet afterimage will reveal to you the same essential truths about particular instances of scarlet. That is surprising, given that, when you hallucinate or have a sensation, you are not encountering any mind-independent instances of scarlet.

The rich pluralist can explain all of this. They say that regardless of whether you are seeing a scarlet pepper, hallucinating a scarlet pepper, or experiencing a red afterimage, you are deeply aware of the *universal* scarlet. This reveals to you part of the essence of *all particular instances* of scarlet for free: it is a priori that all instances of a property share certain aspects of the essence of the property itself. However, you are not deeply aware of *instances* of scarlet. That is why experience does not reveal any differences in essence between such instances, no matter how carefully you look.

Fourth asymmetry You are, let us suppose, very familiar with peppers. You see one and immediately recognize it *as* a pepper. Still, we have seen that your experience does not reveal to you anything about the essence of the high-level property of being a pepper. For all your experience tells you, a pepper is, by its nature, an alien spaceship or an insect cocoon.

Rich pluralism explains this. DA2 says that you are only deeply aware of sensory qualities such as scarlet and pepper-shapedness; you are not deeply aware of the high-level property of being a pepper. At best, you successfully sensorily represent this high-level property. Thus, nothing of its essence is revealed to you.

Summing up and looking forward When we inquired into which essential truths are revealed by experience, we uncovered some surprising asymmetries. Rich pluralism explains them all: it predicts the truth of extended partial revelation.

Can other theories do the same?

3.5 Alternative Explanations of the Asymmetries

In this section, I first argue that several other pluralistic theories can also explain the asymmetries. I then argue that most monistic theories cannot explain the asymmetries but that a certain unusual kind of monistic theory is explanatorily adequate.

Alternative pluralistic explanations I understand *representationalism* as the view that (i) to perceive is to deploy at least one kind of awareness, typically of external particulars and their properties (or property-instances), and (ii) every kind of awareness constitutive of perception is *representational* (i.e., subject to error) in a sense that will be pinned down in chapter 5. Similarly, I understand *naïve realism* as the view that accepts (i) but adds that (iii) every kind of awareness constitutive of perception is *non*representational (i.e., immune to error) in the relevant sense. As far as I am aware, all actual representationalists and naïve realists are monists: they all think that what it is to perceive is to deploy *exactly one* kind of awareness. However, I have carefully constructed the definitions above to leave room for *pluralistic* versions of representationalism and naïve realism—versions on which what it is to perceive is to deploy two (or more) very different kinds of awareness.

Pluralistic representationalists and naïve realists can easily explain the asymmetries about revealed essences: they need only recycle the rich pluralist's explanations. In particular, they can posit both (a) partly essence-revealing awareness of sensory qualities, which occurs in conscious perceptions *and* hallucinations, and (b) non-essence-revealing awareness of particulars, which occurs in perceptions *but not* in hallucinations. Of course, these theorists will disagree with the rich pluralist, and with each other, about which of these kinds of awareness are representational and which are not. But the rich pluralist's explanations, as developed in the previous section, did not turn on these details.

In fact, there is one more pluralistic theory worth mentioning: *strange pluralism*. Strange pluralism posits two kinds of awareness, (a) and (b), just as above, but it says that (a) is representational, while (b) is nonrepresentational. As we will see in later chapters, this theory has little to recommend it over other versions of pluralism. Still, it does have the merit of providing access to the rich pluralist's explanation of extended partial revelation.

I conclude that the asymmetries about revealed essences can be explained by all major pluralistic options: not only rich pluralism but also by pluralistic representationalism, pluralistic naïve realism, and strange pluralism.

Alternative monistic explanations According to the *monist,* what it is for a subject to (consciously) perceive a scarlet pepper is for them to enjoy at most one kind of sensory awareness, whether representational or not. Can the monist explain the asymmetries that we have uncovered?

I believe that standard monistic theories cannot, but some unusual monistic theories can. To defend this conclusion, I will lay out an initial argument against monism based on the asymmetries. The monistic theories that can explain the asymmetries will be the ones that can justifiably reject some step in the argument.

The argument against monism (hence "AM") runs as follows:

AM1. If monism is true, then what it is to perceive is to deploy a single kind of sensory awareness that reveals at least a substantial portion of the essences of its targets.

AM2. It is not the case that what it is to perceive is to deploy a single kind of sensory awareness that reveals at least a substantial portion of the essences of its targets.

AM3. Monism is not true.

In support of AM1, suppose for the sake of argument that monism is true. Then what it is for a subject to perceive a scarlet pepper is for them to deploy at most one kind of sensory awareness. Now, I have argued at length that the subject's sensory awareness of scarlet reveals at least a substantial portion of the essence of scarlet. So, if perceiving is *in general* a matter of deploying this single kind of sensory awareness, then perception should *in general* reveal at least a substantial portion of the essences of its targets.

The case for AM2 is that we perceive peppers, but as I have argued at length, this does not reveal anything about their essences. Thus, perception cannot, in general, be a matter of deploying a single kind of sensory awareness that reveals at least a substantial portion of the essences of its targets.

How might the monist respond?

First consider AM2, which says that it is not the case that what it is to perceive is to deploy a single kind of sensory awareness that reveals at least a substantial portion of the essences of its targets. The support for AM2 was that when you see a scarlet pepper, you are sensorily aware of it, but nothing substantial about its essence is revealed to you. The monist might think to challenge this reasoning by saying that perception does *not* make you sensorily aware of the pepper; it only makes you sensorily aware of various sensory qualities, understood as universals. Since this theory posits a single form of sensory awareness that is *always* revelatory, I will call it *pure revelatory monism*. The representationalist might develop pure revelatory monism by saying that what it is to see a scarlet pepper is to deploy a perceptual representation with a purely **existential content**, say, <there is something scarlet and pepper-shaped in front of me now>. Alternatively, the naïve realist might develop pure revelatory monism by saying that what it is to perceive is to stand in a nonrepresentational relation of sensory awareness to various sensory qualities but not to ordinary objects like peppers.

Pure revelatory monists can explain the asymmetries about revealed essences. They can say that you are not sensorily aware of the pepper or the property of being a pepper: thus, nothing is revealed to you about the essences of these entities. These theorists can add that you stand in a partly revelatory relation to the universal scarlet: thus, a substantial part of its essence gets revealed to you. And, as before, having the essence of the universal scarlet partly revealed to you will automatically reveal to you part of the essences of any instances of scarlet, without revealing to you any essential differences between distinct instances of scarlet.

However, recall the framing assumption from chapter 1. Part of the assumption is that when you see an ovoid orange mango, you are sensorily aware of it and its color and shape, but you are not sensorily aware of any peculiar mental entities. Pure revelatory monism violates this assumption: it entails that perception never makes you *sensorily* aware of particulars like mangoes. (At best, the pure revelatory monist can posit *cognitive* awareness of particular mangoes.) Again, this flies in the face of the phenomenology of perception. It seems introspectively obvious that seeing a mango makes you aware of *that particular mango*, not just abstract color and shape universals, and this awareness seems to be distinctively *sensory*, not cognitive.

Thus, I will set pure revelatory monism aside.

Now return to the other premise of the argument:

AM1. If monism is true, then what it is to perceive is to deploy a single kind of sensory awareness that reveals at least a substantial portion of the essences of its targets.

The monist can sensibly challenge this premise. The reasoning behind the premise was that perception is *sometimes* partly revelatory, so if perceiving is just a matter of deploying a single kind of sensory awareness, then perception should *always* be partly revelatory. To this, the monist might respond that a single, unified kind of awareness—call it *sensing*—might reveal the essences of *some* entities, such as the sensory qualities, without revealing the essences of *other* entities, such as objects and high-level properties. How could this be? Well, perhaps there is something special about the sensory qualities, so that *only they* get essential truths about them partly revealed when sensed.

I will call this view *impure revelatory monism*—*revelatory* because it says that sensing at least sometimes reveals a substantial portion of the essences of its targets, *impure* because it says that this happens for some targets but not others. Impure revelatory monism comes in representationalist and naïve realist varieties, since sensing can be treated as representational or not.[8]

Impure revelatory monists face a crucial question: what is this special feature of the sensory qualities and their instances, the feature that lets *their* essences get revealed when sensed, even while sensing does not reveal the essences of objects and high-level properties? The best answer I can see comes in the form of the following claims ("S" is for "sensing" and "special feature"):

S1. By its nature, sensing reveals, or at least purports to reveal, at least part of how its targets are intrinsically.

S2. It is always a priori that if a sensory quality (or instance thereof) is intrinsically a certain way, then it is *essentially* intrinsically this way.

S3. It is never a priori that if an object, event, or high-level property is intrinsically a certain way, then it is *essentially* intrinsically this way, or even essentially that way at all.

Here, how an entity is *intrinsically* is how that entity and its parts are in and of themselves, regardless of any relations to other entities.

Claims S1–S3 each enjoy at least some independent plausibility. I will make a more systematic case for S1 in chapter 6, but its basic appeal is easy to appreciate: an experience of a scarlet pepper seems to reveal that the pepper is intrinsically scarlet and pepper-shaped and also that scarlet is intrinsically *that way*. S2 is also plausible. Universals might have their (higher-order) *extrinsic* properties accidentally: scarlet might be my favorite property, without that being any part of its essence. But it is hard to see how a universal could have its (higher-order) *intrinsic* properties accidentally. If scarlet intrinsically has a certain saturation, then surely that is essential to it—and this seems to be a priori. Finally, we come to S3. This, too, is very plausible. Consider a particular pepper: it has some of its intrinsic properties, like the property of having mass, essentially, and others, like the property of being scarlet, accidentally. Which intrinsic properties fall into which category is not an a priori matter. (For more discussion of S1–S3, see appendix 2A of chapter 2.)

So suppose that S1–S3 are true. That yields a principled explanation for why sensing reveals a substantial portion of the essences of the sensory qualities and their instances, *without* revealing anything about the essences of objects or high-level properties. For if S1 is true, then to sense any target is to represent how it is intrinsically. And if S2 is true, then this, by itself, reveals part of the very essences of sensory qualities and, derivatively, of their instances. But given S3, this does *not*, by itself, reveal any part of the essences of objects and high-level properties.

Thus, I will assume that the impure revelatory monist accepts claims S1–S3: individually, these claims are plausible, and jointly, they deliver an elegant explanation of extended partial revelation.

However, I do not see any other escape routes from the argument against monism. I conclude that impure revelatory monism remains in the running, but it is the only monistic theory to enjoy this distinction.

A concern I have argued that various pluralistic theories, as well as impure revelatory monism, can explain extended partial revelation. Some will worry that there is no real *explanation* happening here. To be sure, the theories just listed do not take extended partial revelation as primitive—for example, the rich pluralist only arrives at extended partial revelation by putting together three separate posits, DA1–DA3. But the worry is that the "explanation" is very thin: it is that deep awareness occurs in perception (DA1), targets only the sensory qualities (DA2), and is substantially essence-revealing (DA3). It is hard to find much space between these claims and extended partial revelation itself.

Before I engage with this concern, I offer a clarification: I am using the term *explanation* liberally. I count taking a claim as primitive is a way of *explaining* it—it is the limit case of explanation. Given this terminological point, my opponent's worry is not that the theories listed *do not* explain the asymmetries; it is that the explanations are *very thin.*

I agree. I just do not find this worrying. For I have argued that most monistic theories cannot give *any* explanation of truths like partial revelation, *not even very thin ones*. So if some theories can explain the asymmetries at all, *even thinly*, they are preferable to the alternatives.

Still, it would be *even better* to provide more substantial explanations, perhaps by identifying the mechanisms through which essences get partly revealed in experience. I will not do this. But the project of identifying such mechanisms is not the project of developing new, rival theories. It is the project of filling out the theories that I have already identified.

Summing up; looking forward There are some surprising asymmetries, summed up by extended partial revelation, about which essences are partly revealed to us in experience. In this section, I have argued that these asymmetries can be explained by a wide range of pluralistic theories: not only rich pluralism but also pluralistic representationalism, pluralistic naïve realism, and strange pluralism. But, I have also argued, almost no monistic theories are up to the task—the only exception is impure revelatory monism.

Still, there is a major concern about all of the explanations of extended partial revelation that I have suggested.

3.6 In Defense of Experiential Awareness of Universals

I have argued that extended partial revelation is true *because* experience makes us sensorily aware of certain universals, the sensory qualities. This is common ground for all of the theories that I have recommended. But there is a battery of well-known objections to the idea that experience makes us sensorily aware of universals. In this section, I will offer my replies. These are not always original, but they are worth stating for the record.

Objection When you see a scarlet pepper, you do not seem to be sensorily aware of an airy universal in Plato's heaven. You seem to be sensorily aware of the *particular instance* of scarlet instantiated by the pepper.[9]

Reply First, a small point about talk of "Plato's heaven." What I posit are universals *that need not be spatiotemporally located*. A fortiori, they do not reside in some Platonic heaven. The metaphor may entertain, but do not let it mislead.

Now for the substantive point. I agree that seeing a scarlet pepper makes the subject sensorily aware of a particular instance of scarlet. This plausible idea is folded into all of the theories that I have recommended: they all say that perception makes us sensorily aware of particular instances of the sensory qualities. But it is a further step to say that we are not, *in addition to this*, sensorily aware of the universal scarlet. I see no support for this further step.

But wait—don't I then predict that the perceiver is sensorily aware of *both* the universal scarlet *and, additionally*, an instance of it? Yes. And isn't it implausible to posit two distinct kinds of sensory awareness here? I do not see why. Certainly it is not phenomenologically obvious that seeing a scarlet pepper involves these two distinct kinds of awareness. But that is not yet a problem: it certainly seems possible for distinct kinds of awareness to be woven together in such a way that the subject cannot tell them apart through brief reflection alone.

Well, can I at least substantiate the idea by showing that these two kinds of awareness can come apart? Yes: I will argue in chapter 5 that this happens in unconscious perceptions.

A cluster of objections For sensory experience to make us sensorily aware of an entity, arguably the entity must be spatiotemporally extended *and* spatiotemporally located *and* causally efficacious. Universals have none of these features! That is three strikes against the claim that sensory experience makes us sensorily aware of universals. To regiment these objections ("TA" stands for "threefold argument"):

TA1. All targets of sensory awareness are spatiotemporally extended (/spatiotemporally located/causally efficacious).

TA2. Universals are not spatiotemporally extended (/spatiotemporally located/causally efficacious).

TA3. Universals are not targets of sensory awareness.[10]

Reply These objections are powerful. It will take some work to see where they go awry.

One tempting response is to reject TA2 by taking an *Aristotelian view* of universals. On this view, the universal scarlet is wholly present in any entity that is scarlet; it is located and extended over the surfaces of scarlet objects (or their volumes, for scarlet liquids and the like), and it has causal effects, perhaps upon light or the retina.

Suppose that this Aristotelian view is true. Unfortunately, that would not block a reformulated version of the objections. For, plausibly, the universal scarlet must not only be extended and located *somewhere* and have causal effects on *something* but also must have causal effects *on the sense-organs of the subject who is experiencing it*. It must be *this* universal, the one (perhaps wholly present in some scarlet object) exerting its causal power on the subject, that must be spatiotemporally extended and located. However, a hallucinating or sensorily imagining subject need not be in causal communion with any spatiotemporally extended and located universal scarlet, at least not one that is in the external world: there just may not be any scarlet object around for the subject to interact with.

The reformulated objections may be expressed like this:

TA1*. All targets of sensory awareness are spatiotemporally extended (or spatiotemporally located) entities *that have causal effects on the sense-organs of the subject.*

TA2*. Universals are not spatiotemporally extended (or spatiotemporally located) entities *that have causal effects on the sense-organs of consciously hallucinating or sensorily imagining subjects.*

TA3*. Universals are not targets of sensory awareness *for consciously hallucinating or sensorily imagining subjects.*

TA2* is hard to deny. Thus, regardless of whether an Aristotelian view of universals is true, I must reject TA1* (and, in the original argument, TA1).

Are there any grounds for rejecting TA1*? I think so: they are provided by cases of sensory imagination. Suppose that you vividly imagine a scarlet fruit and a mango orange one, where these are not any particular fruits that you have seen before. Your imaginative experience will provide you with a particular kind of *intelligible basis* for drawing conclusions about the essences of scarlet and mango orange. It is hard to see how you could have access to the specific kind of intelligible basis that you do, unless your experience gave you sensory awareness of scarlet and mango orange. But neither any instances of these colors, nor the universals themselves, have any current causal effects on your sensory organs. Thus, it is at least reasonable to reject TA1*: it is reasonable to say that you are sensorily aware of universals that have no current effects on your sense-organs.

We are not out of the woods yet. We still face a pressing question: How is it even *possible* to be sensorily aware of universals that are not currently having any effect on your sense-organs? What could the underlying mechanism be?

I do not know, but here are some possible answers. The dualist might say that there is no underlying mechanism. It is just a brute law of nature that when certain neural states occur, there is deep awareness of certain universals.[11] By contrast, the physicalist might identify a proper mechanism. They might say that my ancestors encountered *instances* of, for example, scarlet, which were of course spatiotemporally located, spatiotemporally extended, and causally efficacious. (They would thereby incur the burden of explaining away the appearance, which I discuss in the next chapter, that colors could not possibly be instantiated in a purely physical world.) The physicalist might add that my ancestors thereby developed the capacity to be sensorily aware of the *universal* scarlet and could exercise that capacity even when they were not causally interacting with instances of scarlet. Because I have a capacity with this evolutionary history, I, too, can be sensorily aware of the universal scarlet even if I have *never* encountered an instance of scarlet.[12]

These are just two possible accounts. I am open to others.

I want to conclude with a diagnosis of why this cluster of objections seems so compelling. I suggest that it is because my opponent has correctly identified some requirements on experiential awareness of a *particular*. As I will argue in chapter 5, this *does* require the subject to be in causal commerce with the particular, which cannot happen if the particular is not spatiotemporally located and causally efficacious.[13] But we should not confuse requirements on experiential awareness of particulars with requirements on experiential awareness of universals.

Objection It is plausible that if you see an entity, then that entity at least *looks* spatially extended to you (even if it is not *really* spatially extended). Universals do not look spatially extended. Hence, we do not see them.[14]

Why accept that universals do not look spatially extended to us? My opponent provides three supporting arguments here. First, universals are *necessarily* unextended, so it is hard to see how they could look extended. Second, if scarlet looks extended in cases of hallucination, then it also looks extended in cases of perception. But then when I see a scarlet pepper, there would be two entities that look extended to me: the pepper *and* the universal scarlet. That is evidently not the case. Finally, if an entity looks extended, then it looks to have a shape. But scarlet does not look to have a shape.

Reply The conclusion of this objection is that we do not see universals. The extended partial revelationist is welcome to accept this conclusion. For instance, the rich pluralist says that to (consciously and veridically) see a scarlet pepper is (i) to successfully sensorily represent the pepper and its instances of scarlet, (ii) to be deeply aware of the universal scarlet, and

(iii) to represent these instances of scarlet *as* instances of the universal scarlet. On one version of this theory, the only entities that we can successfully sensorily represent—and thus the only entities that we can see—are particulars such as objects and property-instances. On another version of the rich pluralist's theory, we represent particular property-instances *as* instances of certain universals, but this does not involve *seeing* universals. It only involves seeing particulars and representing them as falling under universals. Either way, universals cannot be seen. Nor can they look any way to us, since only something that is seen can have a look.

An analogy will clarify the basic strategy. Lance is riding a bicycle, and for the bicycle to be ridden, it must have wheels. But what is ridden is the bicycle itself, not the wheels. Similarly, we consciously see objects and property-instances, which entails that we have deep awareness of universals. But what are seen are the particulars, not the universals.[15]

Revised objection Grant that universals cannot be seen. Still, it is easy enough to slide a similar objection through the bars. Just appeal to the principle that if you are *visually aware* of an entity, then that entity at least *visually seems* spatially extended to you. Then point to more or less the same three reasons as before for thinking that universals do not visually seem to be spatially extended: first, that universals are necessarily not extended; second, that being visually aware of a pepper does not involve being visually aware of an extended universal, scarlet, in addition to the extended pepper; and third, that entities that visually seem to be extended also visually seem to have a shape, but scarlet does not visually seem to have a shape.

Reply This objection gets to the heart of the matter. My response is to deny that if you are visually aware of an entity, then that entity visually seems to you to be spatially extended. For suppose that I veridically see a ball and attend to its shape. Set aside universals and focus on property-*instances*: surely I am visually aware of an *instance* of the ball's sphericity. But this instance of sphericity is not *itself* spatially extended. If it were, it would have a shape, but it is the spherical *object*, not the *instance of sphericity*, that is spherical. Thus, if we accept the principle at issue, then we must say that the instance of sphericity visually seems to me to be spatially extended despite essentially failing to be spatially extended. This is a strong reason to reject the principle.[16]

Summary It is often thought that the ontological nature of universals precludes any awareness of them in experience. On reflection, I believe that this objection is not well supported.

3.7 Conclusion

It is odd what gets revealed to us in experience. All sensory experiences reveal *some* essential truths to us. These truths are especially substantial, though not complete, for sensory qualities; they are somewhat more restricted for instances of sensory qualities; and they are missing entirely for objects, high-level properties, and instances of high-level properties. To weave all of this together:

> *Extended partial revelation*: In the first instance, all sensory experiences reveal a substantial portion of the essential truths about the sensory qualities and nothing else.

This is an important result: extended partial revelation rules out almost all monistic theories, whether representationalist or naïve realist. For monists posit just a single kind of sensory awareness. If that awareness is taken to be substantially essence-revealing, then it is natural for the monist to predict that seeing a pepper will reveal a substantial part of what it is to be a pepper. If that awareness is *not* taken to be substantially essence-revealing, then it is natural for the monist to predict that seeing scarlet will not reveal any part of what it is to be scarlet. Both predictions are false.

We have discovered several pluralistic theories, as well as one monistic theory, that can do better. The pluralistic theories all posit one kind of awareness of universals, which is partly essence-revealing, together with a second kind awareness of particulars that is not at all essence-revealing. The monistic theory, *impure revelatory monism*, posits a single kind of awareness, *sensing*, that is partly essence-revealing for the sensory qualities but not for particulars. I conclude that we should select one of these theories.

The next chapter will have many tasks, but a central one will be to support extended partial revelation with another argument, one focusing on asymmetries regarding certain *hard problems*. Only in later chapters will I argue for rich pluralism over all other theories that accept extended partial revelation.

4 Hard Problems of Consciousness and the External World

There are many hard problems.[1] But some of them—in particular, those concerning *consciousness* and the *external world*—have an especially interesting connection to experience and are the topic of this chapter. More precisely, this chapter has three tasks.

The first task is to explain when these hard problems arise. The task is not to *solve* the hard problems; I have no particular solution in mind. The task is rather to explain why certain questions constitute hard problems for us, while other, very similar questions do not. I show that existing theories are not up to this task, and I develop a new theory, the **E-theory**, that can do better (sections 4.1–4.4).

The second task of this chapter is to provide new lines of support for extended partial revelation. I do this by arguing that extended partial revelation receives substantial support from the E-theory (sections 4.2 and 4.5).

The third and final task of this chapter is to identify, and provide a substantive theory of, a form of consciousness that is distinctively associated with the hard problems (sections 4.6–4.8).

4.1 Data about the Hard Problems of the External World

In this section, I consider a series of questions about the *external world*. I argue that some of these questions constitute hard problems, even while other, very similar questions do not. This makes for some puzzling asymmetries. The next section attempts to explain why these asymmetries occur.

We should start by understanding what a hard problem is. Let us say that for the fact that *p* to *partly/fully ground* the fact that *q* is for *q* to hold partly/fully in virtue of *p*, metaphysically speaking. The idea is that **grounding** is a relation between facts, or perhaps states of affairs, that backs metaphysical explanations in something like the way that causation backs causal explanations.[2] To give some potential examples: the existence of this table is arguably fully grounded in the existence of certain particles that were arranged tablewise by beings with certain intentions, the existence of a triangle is arguably fully grounded in the existence of a three-sided closed planar figure, the existence of {Socrates} is arguably at least partly grounded in the existence of Socrates, and the disjunctive fact that *p* or *q* is arguably fully grounded in the true disjunct(s).

We can then say that for there to be a *hard problem* is for the following to be true (where *p*, *q*, and *r* range over propositions):

(i) we have strong background evidence that *p* fully grounds *q*—but, at the same time,

(ii) it strongly initially seems impossible to us that *p* fully grounds *q*: we cannot see why *p* would make it the case that *q* rather than making it the case that *r*.

When these conditions are met, I will also say that the question, "Why is it the case that *q* rather than *r*?" *constitutes* a hard problem.

It will be easier to understand this definition once we have seen it in action. Let us consider some data about which external-world questions constitute hard problems and which do not.

First datum Assume that perception is not massively illusory. Given this assumption, some peppers are scarlet, some violins produce notes of middle C, and some puddings are vanilla-scented. But it is very hard to see how any of this could be explained in purely physical terms. Consider, for instance, this question about a particular scarlet pepper ("EQ" stands for "external-world question"):

EQ1. Why is the pepper scarlet rather than lime green?

This question constitutes a hard problem. For, on the one hand, we have strong scientific evidence that the pepper's being scarlet is fully grounded in its having various surface reflectance properties. But, on the other hand, this strongly initially seems to be impossible. Even after digesting the latest scientific theories, it is hard to see how the pepper's having certain surface reflectance properties could *make it the case* that the pepper is scarlet. Being scarlet just does not seem to be a matter of having surface reflectance properties. The point generalizes to other sensory qualities. How could mere vibrations in the air *make it the case* that there is an instance of middle C? How could mere chemical properties *make it the case* that a pudding is vanilla-scented?[3]

Notice that EQ1 is not a question about experience. It is purely a question about the external world. The question would be just as hard to answer if we were asking about scarlet peppers, middle-C sounds, and vanilla-scented puddings in a world that was entirely free of conscious subjects.

Second datum Look again at a scarlet pepper, and reflect on a different question:

EQ2. Why does the pepper exist rather than not existing?

I submit that this question does not constitute a hard problem: there is no particular obstacle to understanding how the existence of *that particular object* could be fully grounded in the physical facts. There is an obstacle to understanding how the pepper could be *scarlet*, but that is a separate matter from its existence.

Against this, some have argued that EQ2 *does* constitute a hard problem. In particular, it has been said that the pepper is a *mereological whole* and that it is hard to see how the existence of a whole could be fully grounded in the existence of its physical parts.[4]

I do not find this line of thought compelling. I think that there are just different concepts of wholes, and *once we fix on a particular concept*, there is no hint of any hard problem about mereology. For instance, we might, in the spirit of mereological universalism, take it as a conceptual truth that a whole *just is* its parts. Then there is certainly no problem in understanding why the existence of a whole pepper is fully grounded in the existence of its parts. Or, if we are in a more discerning mood, we might stipulate that a whole is a collection of parts that are *unified* in some special way. Then, once we specify what kind of unity we have in mind, we can check whether the pepper is a whole by checking whether its parts are unified in that way.

This approach certainly leaves us with plenty of difficult philosophical questions about wholes. For instance: which concepts of wholes are used by the folk in various contexts? And which concepts of wholes are the most fruitful for various philosophical purposes? But none of these questions constitute hard problems, in the present sense.

Third datum Consider:

EQ3. Why does this pepper instantiate *this* instance of scarlet rather than instantiating some other instance of scarlet?

EQ3 does not constitute a hard problem. It is hard to understand why the pepper is scarlet rather than lime green, but there is no great mystery in understanding why the pepper instantiates this instance of scarlet rather than the instance of scarlet found in a nearby apple. The apple's instance of scarlet is just in the wrong place.

Fourth datum Suppose that you see a pepper and recognize it as such. You then learn the botanical account of what a pepper is. Afterward, you look at the pepper and ask yourself:

EQ4. Why is this a pepper rather than a pine tree?

EQ4 does not constitute a hard problem: you will have no trouble squaring your scientific knowledge of peppers with what you see before you. You might have trouble understanding how the pepper could be scarlet, but you will have no trouble understanding how it could be a pepper or how a pepper could *just be* the fruit of a particular kind of plant. More generally, there are no hard problems associated with instances of high-level properties.

Fifth and six data Consider the following questions:

EQ5. Why is the pepper pepper-shaped rather than spherical?

EQ6. Why is the pepper still rather than moving?

Neither of these questions constitute hard problems. There is no particular difficulty in seeing how purely physical facts could fully ground the existence of instances of shape properties, motion properties, or other so-called primary qualities.

Conclusion There are hard problems about why objects instantiate one sensory quality (say, scarlet) rather than another (say, lime green). But there are no hard problems about why objects exist, why they instantiate one instance of a sensory quality rather than another instance of the same quality, or why they instantiate high-level properties. Nor are there hard problems about why objects instantiate primary qualities like pepper-shapedness or stillness.

These are strange patterns. How should we explain them?

4.2 Explanations via Extended Partial Revelation

Recall this thesis, which I defended carefully in previous chapters:

> *Extended partial revelation*: In the first instance, sensory experiences reveal a substantial portion of the essences of sensory qualities and of nothing else.

In this section, I will argue that by conjoining extended partial revelation with a single additional claim, we can explain the data about the hard problems of the external world. This is another advantage of accepting extended partial revelation.

The explanation I have in mind is this (it will become clear later in the chapter why I call this thesis "E1"):

E1. Often, it strongly initially seems impossible to us for purely physical entities to satisfy the essential conditions that are revealed to us in sensory experience (as described in extended partial revelation). When and because this happens, there are hard problems.

Let us see how extended partial revelation and E1 can explain the data that we have found.

Start with:

EQ1. Why is the pepper scarlet rather than lime green?

Our first datum was that EQ1 constitutes a hard problem. We can now explain why. According to extended partial revelation, when I perceive a scarlet pepper, a substantial portion of the essence of scarlet is thereby laid bare to me—for example, it is revealed to me that it lies in the essence of scarlet, and hence all of its instances, to have a certain intrinsic character, call it *I*. E1 adds that we find it hard to see how purely physical entities could satisfy this essential condition. For example, we find it hard to see how the pepper's having various surface reflectance properties could fully ground its having a property with intrinsic character *I*. According to E1, that generates a hard problem.

Now recall:

EQ2. Why does the pepper exist rather than not existing?

Our second datum was that EQ2 does not constitute a hard problem. The present account lets us explain why not: it is because experience does not reveal to us any essential truths about ordinary objects (as per extended partial revelation). Thus, there is no possibility that we might have trouble seeing how those essential conditions could be met by purely physical entities (as per E1).

Next, consider:

EQ3. Why does this pepper instantiate *this* instance of scarlet rather than some other instance?

Our third datum was that EQ3 does not constitute a hard problem. The explanation is that, in the first instance, experience reveals substantial essential truths only about the sensory qualities; it reveals nothing about what distinguishes different instances of the same sensory quality (extended partial revelation). But EQ3 concerns different instances of the same sensory quality. Thus, no hard problems will be generated (E1).

The next question on the agenda is:

EQ4. Why is this a pepper rather than a pine tree?

This does not constitute a hard problem. The current account explains why not: experience never reveals anything about the essences of high-level properties (extended partial revelation), so there is no opportunity for any hard problems to arise about them (E1).

Finally, return to these questions:

EQ5. Why is the pepper pepper-shaped rather than spherical?

EQ6. Why is the pepper still rather than moving?

The fifth and sixth data were that neither of these questions constitute hard problems. Why not? We can explain this as follows. Yes, when we see peppers in ordinary conditions, a substantial part of the essence of pepper-shapedness, perhaps even all of it, is laid bare to us. Likewise for the essence of stillness (extended partial revelation). But it is easy to see how purely physical entities could, when arranged in certain ways, meet those essential conditions. Thus, no hard problems arise in these cases (E1). More generally, E1 does not predict that *every* sensory quality will be associated with a hard problem of the external world. It only predicts that hard problems will arise *when we find it hard to understand how the revealed essential conditions can be met by purely physical entities*.

This last suggestion might raise a red flag. I am saying that some sensory qualities are associated with hard problems, and others are not, because sometimes we find it hard to see how the revealed essential conditions can be met physically, and other times we do not. But if we cannot independently tell *which* sensory qualities fall into which of these two categories, then my account loses much of its explanatory power.

Fortunately, we *can* independently tell which sensory qualities fall into which category. I am not just positing, without evidence, that we can readily understand how the essence of pepper-shapedness can be realized physically. Rather, it is *independently obvious* that this claim is true. It is independently obvious that once we understand the mathematical description of pepper-shapedness, we will readily understand why that *is* the correct mathematical description of the shape property we experience when we see the pepper. And it is no trouble to see how physical particles could be arranged so as to satisfy that mathematical description. More generally, we can independently tell whether it will seem possible, or impossible, for a revealed essential condition to be met by a purely physical entity. Thus, my account really does *informatively explain* why scarlet is associated with a hard problem and why pepper-shapedness is not.

Conclusion There are complex data about which questions do, and which questions do not, constitute hard problems of the external world. E1, together with extended partial revelation, explains all of these data. If others have advanced competing explanations of these data, I am not aware of them. This is another powerful reason to accept extended partial revelation.

4.3 Data about the Hard Problems of Consciousness; Some Failed Theories

So far, we have been considering the hard problems of the external world. This section turns to the hard problems of consciousness. I begin by explaining what a hard problem of consciousness is. I go on to identify some curiously asymmetrical data about when such problems arise, and I show that these asymmetries undermine a series of influential theories about what generates the hard problems of consciousness.

What is a hard problem of consciousness? I have already said what it is for there to be a *hard problem* in general. It is for the following to be true (where *p*, *q*, and *r* range over propositions):

(i) we have strong background evidence that *p* fully grounds *q*—but, at the same time,

(ii) it strongly initially seems impossible to us that *p* fully grounds *q*—we cannot see why *p* would make it the case that *q* rather than making it the case that *r*.

I now stipulate that a *hard problem of consciousness* is simply a hard problem in which *q* is a proposition about the existence, or specific character, of one or more conscious experiences. To take a paradigmatic example, suppose that you see a scarlet pepper. It strongly initially seems impossible for the existence of your experience to be fully grounded in physical facts alone. Whatever you learn about neural firing rates and surface reflectance properties, it will remain unclear why that would make it the case that you are having that particular *conscious experience.*

My definition of a hard problem of consciousness is slightly different from what is usual. It is more usual to begin by identifying particular paradigms of hard problems—for example, problems about explanatory gaps, the possibility of knowing all the physical facts without knowing all the consciousness facts, and the rational conceivability of worlds physically identical to ours but containing different conscious experiences or none at all. A "hard problem of consciousness" is then said to be any problem that is appropriately similar to these paradigmatic problems. In section 4.7, I will make it clear why I prefer the current terminology: I will argue that hard problems, in my sense, are defined by the underlying feature that generates all of the particular "hard problems," in the usual sense. In the meantime, I will just use my own definition.

Now that we understand what a hard problem of consciousness is, we can ask: what particular hard problems of consciousness are there?

Data about some hard problems of consciousness and potential theories to explain them The simplest possible theory of the hard problems of consciousness is this:

> The *A-theory*: There are no hard problems of consciousness.

(I am just unimaginatively labeling these theories starting from the beginning of the alphabet. I am *not* using the well-known lettering system devised by Chalmers.)

The A-theory is hard to sustain. Look at a scarlet pepper, and consider these questions about your experience, where "CQ" stands for "consciousness question":

CQ1. Why do you experience scarlet rather than not experiencing anything?

CQ2. Why do you experience scarlet rather than experiencing lime green?

Intuitively, both of these questions constitute hard problems of consciousness: it is difficult to see how any answer to them could be fully grounded in the physical facts alone. Facts about brain processes and surface reflectance properties just do not seem to provide a complete answer to either question.[5] Regarding CQ1, it is mysterious why light bouncing off a pepper, entering the retina, and getting processed in a certain way should generate an *experience*—why it would generate *consciousness*—rather than mere unconscious information processing. Regarding CQ2, grant for the sake of argument that we understand why you are having an experience in the first place. It remains mysterious why you are having an experience of *scarlet* in particular. Even once we understand all of the relevant properties of the pepper and of your brain, it is hard to see where scarlet comes into things.

Some have tried to defend the A-theory from this attack: they have claimed that naïve realism, in particular, gives us a completely satisfying way of grounding the answers to CQ1 and CQ2 in physical facts alone. In answer to CQ1, these theorists say that what it is to perceive is to stand in a relation of *acquaintance*—an irreducible relation that can only be directed at properties instantiated in the subject's immediate environment. It is in virtue of being acquainted with something that the subject has an experience in the first place. In answer to CQ2, they say that some peppers are scarlet. It is because you are acquainted with an instance of scarlet rather than an instance of lime green that you experience scarlet in particular. It is perfectly transparent why being acquainted with something would generate an experience and why being acquainted with scarlet in particular would generate an experience of scarlet rather than lime green. Thus, CQ1 and CQ2 do not really constitute hard problems.[6]

I do not think this is a good defense of the A-theory.

Regarding CQ1, I grant that it is perfectly clear why being acquainted with something would generate an experience: we are just stipulating that acquaintance is the kind of relation that is sufficient for conscious experience. But if acquaintance is being understood in this rich way, then *how on earth do purely physical processes fully ground the existence of acquaintance?* It is a mystery. Indeed, it is pretty nearly the same mystery as that of understanding how purely physical processes could fully ground the existence of an experience. Thus, CQ1 still constitutes a hard problem.

Regarding CQ2, I grant that it is perfectly clear why being acquainted with scarlet would generate an experience of scarlet rather than lime green. But how on earth is it possible for a purely physical object to be scarlet in the first place? What do physical properties, like surface reflectance properties, have to do with *scarlet*, the property that we encounter in experience? This, as I observed in the previous section, is another mystery. Thus, CQ2 still constitutes a hard problem.[7]

We really do need to explain, not just explain away, the appearance that CQ1 and CQ2 constitute hard problems. How can we do so? Well, we saw in the previous section that there are hard problems of the external world. Some thinkers have suggested that this suffices to explain the existence of the hard problems of consciousness: if we find it hard to understand how sensory qualities can be instantiated in the first place, then of course we will find it hard to understand how it is possible to *experience* sensory qualities. More precisely, these thinkers accept:

The *B-theory*:

B1. It strongly initially seems impossible to us for purely physical facts to make it the case that any sensory quality is instantiated. When and because this happens, there are hard problems.

B2. There are no other sources of hard problems of consciousness or the external world.[8]

The B-theorist can easily explain why CQ2—the question of why you experience scarlet rather than lime green—constitutes a hard problem. The explanation is that we find it hard to understand how purely physical facts could make it the case that anything *is* scarlet rather than lime green (B1). Thus, we find it hard to understand how purely physical facts could make it the case that we *experience* scarlet rather than lime green.

Now return to CQ1, the question of why you experience scarlet rather than not experiencing anything—rather than having mere unconscious information processing. The B-theorist must give precisely the same explanation for why this question constitutes a hard problem: they must say that we find it hard to understand how purely physical facts could make it the case that anything *is* scarlet in the first place (B1), so we find it hard to understand how purely physical facts could make it the case that we *consciously experience* scarlet.

I find this explanation problematic. For it commits the B-theorist to saying that *if we understood how purely physical facts could make it the case that something was scarlet*, then CQ1 would not constitute a hard problem. But this prediction is clearly incorrect. Even if we understood exactly how it was possible for a purely physical pepper to be *scarlet*, we would still find it hard to understand how a purely physical being, standing in a purely physical relation to a genuinely scarlet object, could *experience* that color. Why wouldn't it be all dark inside, so to speak?

We can see the problem even more clearly if we focus on shape experience. There is no need to consider hypotheticals here: we *really do* understand perfectly well how a purely physical object could have the precise shape that the pepper has. Still, it remains mysterious how it is possible for a purely physical being, in a purely physical world, to *consciously experience* that shape.[9] The lesson is that *some* hard problems of consciousness, such as the one constituted by CQ1, are not just about *which particular property* that is being experienced. They are, at least in part, about *the possibility of having an experience at all*. The B-theory is no good.

Well then, perhaps there are two distinct sources of hard problems: one having to do with the sensory qualities, the other having to do with conscious experience itself. Some thinkers have developed these ideas into the following theory:

The *C-theory*:

C1. What it is to consciously perceive is to stand in a single, special kind of awareness relation, the relation of *acquaintance*, to particulars and universals.

C2. It strongly initially seems impossible to us for purely physical facts to make it the case that a subject is acquainted with something rather than not being acquainted with anything. When and because this happens, there are hard problems.

C3. It strongly initially seems impossible to us for purely physical facts to make it the case that a subject is acquainted with one target rather than another. When and because this happens, there are hard problems.[10]

The C-theory can explain both of the data mentioned so far. The first datum was that CQ1—the question, Why do you experience scarlet rather than not experiencing anything?—constitutes a hard problem. The C-theorist can explain this by saying that what it is for you to experience scarlet is for you to be acquainted with scarlet (C1), and we find it hard to understand how purely physical facts could make it the case that there is an acquaintance relation (C2). This avoids the problem that brought down the B-theory: the explanation remains in place even assuming that you understand how it is possible for a purely physical object to be scarlet.

The C-theorist can also explain the second datum, which was that CQ2—the question, Why do you experience scarlet rather than experiencing lime green?—also constitutes a hard problem. This time, the explanation is that what it is for you to experience scarlet is for you to be acquainted with scarlet (C1), and we find it hard to understand how purely physical facts could make it the case that you are acquainted with any given target rather than something else (C3).

The C-theory has a further advantage. Consider this question:

CQ3. Why do you experience this particular pepper rather than not experiencing anything?

Like CQ1 and CQ2, this question constitutes a hard problem. Collect all of the physical facts about your brain, the pepper, and the interaction between the two. None of these facts seem to fully explain why you are having a *conscious experience* of that particular pepper. The C-theorist can explain why CQ3 constitutes a hard problem: they say that for you to experience the pepper is for you to be acquainted with it (C1), and we find it hard to understand how purely physical facts could make it the case that there is an acquaintance relation at all (C2).

Nevertheless, I believe that the C-theory runs into trouble. Consider:

CQ4. Why do you experience this pepper rather than experiencing an intrinsically identical pepper somewhere in Vietnam?

The C-theorist predicts that this question will constitute a hard problem: C1 tells us that for you to experience the pepper is for you to be acquainted with it, and C3 adds that we find it hard to understand how purely physical facts could make it the case that you are acquainted with any given thing rather than some other thing. However, I submit that there is not the slightest glimmer of a hard problem here. CQ4 presupposes that you are experiencing something. Once you know the relevant physical facts—that is light *from this particular pepper* that is entering your retina, that it is *this particular pepper* that your visual system tracks as you move closer to get a better view, and so on—you can understand exactly why it is *this* pepper, not the Vietnamese one, or for that matter a Kenyan elephant, that you are experiencing. For this reason, I reject the C-theory.

We are left with a curious pattern. CQ1 constitutes a hard problem about experienced sensory qualities, and CQ3 constitutes a parallel hard problem about experienced objects: so far, so symmetrical. But whereas CQ2 constitutes another hard problem about experienced sensory qualities, there is no hard problem constituted by CQ4, the parallel question about experienced objects. How odd!

We might try to explain this asymmetry by tweaking the C-theory. We can keep the idea that hard problems get generated partly because we find it hard to understand how purely physical facts could make it the case that there is an experience at all. But we can add, contrary to C3, that hard problems do not get generated by *every* potential target of experience; they only get generated by experienced *properties*.

One version of this theory appeals to property-*instances* rather than universals:

The *D-theory*:

D1. It strongly initially seems impossible to us for purely physical facts to make it the case that a subject has a conscious experience rather than not having an experience at all. When and because this happens, there are hard problems.

D2. It strongly initially seems impossible to us for purely physical facts to make it the case that a subject has a conscious experience of one property-instance rather than another property-instance. When and because this happens, there are hard problems.

D3. There are no other sources of hard problems of consciousness or the external world.

The D-theory explains all of the data discussed so far. It explains why CQ1, the question about why you experience scarlet rather than not experiencing anything, and CQ3, the question about why you experience this particular pepper rather than not experiencing anything, both constitute hard problems. This occurs because a hard problem is constituted by *any* question about why a subject has a conscious experience rather than not having an experience at all (D1).

The D-theory also explains the asymmetry between CQ2 and CQ4. CQ2 asks why you experience scarlet rather than experiencing lime green. The D-theorist correctly predicts that this will constitute a hard problem: the question is about experiences of different property-instances, and according to D2, we find it hard to understand how purely physical facts could make it the case that you experience one property-instance rather than another. By contrast, CQ4 asks why you experience this pepper rather than experiencing a different one. This question is not about why you have an experience rather than not having an experience, nor is it about experiences of different property-instances. But D3 says that these are the only sources of hard problems. Thus, the D-theorist correctly predicts that CQ4 will *not* constitute a hard problem.

To further test the D-theory, let us consider some questions that are specifically about property-*instances*. We can start with this question:

CQ5. Why do you perceive an instance of scarlet rather than perceiving an instance of lime green?

This question obviously constitutes a hard problem of consciousness. The physical facts about your brain, as well as the properties of the pepper, leave it mysterious why it is an instance of *scarlet* that you experience. The D-theory correctly predicts this: D2 tells us that hard problems will be constituted by questions about why you experience one property-instance rather than another.

But then there is this question:

CQ6. Why do you perceive *this* instance of scarlet rather than perceiving some *other* instance of scarlet?

Since CQ6 is about experiencing different property-instances, the D-theory, in particular D2, predicts that the question will constitute a hard problem. This prediction is false. Assume for the sake of argument that you understand why peppers can instantiate scarlet in the first place. (This much, we may suppose, is a presupposition of CQ6.) Then there is no particular mystery about why you might perceive *this* particular instance of scarlet rather than an instance of scarlet on a pepper across the globe. Such facts could be fully grounded in facts about the physical relationship between your brain and the relevant instance of scarlet.

We have found another curious asymmetry: questions like CQ5 constitute hard problems, but questions like CQ6 do not. The D-theory fails to explain this.

Conclusion We want to explain why certain questions constitute hard problems of consciousness, while other questions do not. I have examined three existing theories (the A-, B-, and C-theories) and one new theory (the D-theory). All were found wanting.

4.4 A Better Theory, and Some Further Data

In this section, I construct a new theory, the E-theory (although it will take us a while to get to the final version of the theory). I show that the E-theory can explain all of the data we have seen so far, as well as some new data.

The basic idea of the E-theory is that hard problems are generated, not by questions about different *property-instances* but by questions about different *properties*. Here is an initial sketch:

The *E-theory* (first pass):

E1**. It strongly initially seems impossible to us for purely physical facts to make it the case that a subject has a conscious experience of one property rather than another property. When and because this happens, there are hard problems.

E2**. It strongly initially seems impossible to us for purely physical facts to make it the case that a subject has a conscious experience rather than not having an experience at all. When and because this happens, there are hard problems.

E3**. There are no other sources of hard problems of consciousness or the external world.

You can think of "E" as standing for both *extended partial revelation* and *essences*. The asterisks (e.g., "E1**") will be dropped as I refine the theory, with the final version being asterisk-free.

Like the D-theory, the E-theory correctly predicts the data about CQ1–CQ4. The explanations run just as before, so I will not repeat them. The advantage of the E-theory is that it also explains the asymmetry between CQ5 and CQ6. CQ5 asks why you perceive an instance of scarlet rather than perceiving an instance of lime green. This is a question about experiences of different properties, so E1** correctly predicts that this question will constitute a hard problem. By contrast, CQ6 asks why you perceive *this* instance of scarlet rather than *some other* instance of scarlet. This is not a question about why you experience one property rather than a different property; in both cases, you are experiencing the same property, scarlet. So E1** does not apply. Nor is it a question about why you have an experience in the first place, so E2** does not apply. E3** says that there are no further sources of hard problems. Thus, the E-theorist correctly predicts that CQ6 will not constitute a hard problem of consciousness.

But there are more data to be explained. Suppose that you are familiar with how peppers look, so you experience the scarlet pepper *as a pepper*. Now consider:

CQ7. Why do you experience that object as a pepper rather than not experiencing anything at all?

CQ7 constitutes a hard problem: it is utterly mysterious why, when you look at a pepper, you have a technicolor conscious experience of it as a pepper. The E-theorist can explain why this mystifies us. They say that we have trouble understanding how purely physical facts could ground the existence of a conscious experience in the first place (E2**). A fortiori, we will have trouble understanding how purely physical facts could ground the existence of this particular kind of conscious experience, an experience of something *as a pepper*.

But here is an interesting contrast. Let a *twin-pepper* be a fruit that grows on some alien planet. Twin-peppers look just like peppers, but they are botanically nothing like them: they are to peppers as the water-like substance on Twin Earth is to water. Now reflect on:

CQ8. Why do you experience that object as a pepper rather than experiencing it as a twin-pepper?

Unlike CQ7, CQ8 does not constitute a hard problem. In light of the fact that you have had causal commerce with peppers and not twin-peppers, it is perfectly obvious why you experience the pepper *as a pepper*.[11] Can the E-theorist explain why CQ8 does not constitute a hard problem?

It depends. The key point is that according to E1**, hard problems will be constituted by questions about why we experience one property rather than another. So suppose that the E-theorist says that we experience high-level properties, like the property of being a pepper. Then they end up with the wrong prediction: they wrongly predict that CQ8—the question, Why do you experience that object as a pepper rather than experiencing it as a twin-pepper?—will constitute a hard problem.

The E-theorist can avoid this problem by accepting this further commitment:

E4**. We never experience high-level properties. We only ever experience sensory qualities.

With E4** in hand, the E-theorist gets things right. Recall E1**, the claim that it strongly initially seems impossible to us for purely physical facts to make it the case that a subject experiences one property rather than another. This claim only "triggers" for properties that we experience, and E4** tells us that we never experience high-level properties. Thus, the E-theorist now correctly predicts that CQ8 will not constitute a hard problem.

In sum, the E-theory, if it is supplemented with E4**, seems to explain all of the data about the hard problems of consciousness. It is the best theory we have found so far.

In fact, we can draw a broader lesson. Our current theory explains the data by identifying two core sources of hard problems. First, we find it hard to see how purely physical facts could ground the existence of a conscious experience in the first place. Second, we find it hard to see how purely physical facts could ground our experiencing one sensory quality rather than another. Thus, we can explain all the data by adopting this more general theory:

The *E-theory* (second pass):

E1*. It strongly initially seems impossible to us for purely physical facts to make it the case that a subject has a conscious experience of one sensory quality rather than another. When and because this happens, there are hard problems.

E2*. It strongly initially seems impossible to us for purely physical facts to make it the case that a subject has a conscious experience rather than not having an experience at all. When and because this happens, there are hard problems.

E3*. There are no other sources of hard problems of consciousness or the external world.

The earlier version of the E-theory is just a particular instance of this second version.

The second version of the E-theory explains all of the data so far. Return to the questions of why you experience *scarlet* rather than not experiencing anything (CQ1), why you experience *this particular pepper* rather than not experiencing anything (CQ3), and why you experience this particular pepper *as a pepper* rather than not experiencing anything (CQ7). Since these are all questions about why you have an experience in the first place, E2* correctly predicts that these will all constitute hard problems. Next, consider the questions of why you experience scarlet rather than experiencing lime green (CQ2) and why you experience an *instance* of scarlet rather than an *instance* of lime green (CQ5). Since these are questions about experiencing one sensory quality rather than another, E1* correctly predicts that they will both constitute hard problems. Finally, consider the questions of why you experience this pepper rather than experiencing an intrinsically identical pepper somewhere in Vietnam (CQ4), why you perceive this instance of scarlet rather than perceiving some other instance of scarlet (CQ6), and why you experience the pepper as a pepper rather than a twin-pepper (CQ8). These questions are not about why you experience one sensory quality rather than another, so E1* does not trigger. Nor are they about why you have an experience at all; thus, E2* does not trigger. E3* tells us that there are no additional sources of hard problems of consciousness or the external world. Thus, the current version of the E-theory correctly predicts that these questions will not constitute hard problems.

In fact, the data can be extended further. The hard problems that we have considered so far have all been about *perceptions*. But hard problems equally arise for *all* sensory experiences, including hallucinations, sensations, sensory imaginings, and episodic memories. The physical facts seem to leave it equally mysterious why you consciously *hallucinate* or *imagine* scarlet rather than lime green (or rather than not having an experience at all). The E-theorist can easily explain why that is, since E1*–E3* all concern conscious experiences quite generally, not just perceptions.

So far, then, the second version of the E-theory is the simplest theory that can explain all the data.

Some concerns At this point, some will worry that I have overlooked some important categories of hard problems. First, consider:

CQ9. Why does your perception of the scarlet pepper position you to know *what the pepper is*, in a certain special sense, rather than not positioning you to know this?[12]

You might worry that CQ9 constitutes a hard problem and that the current version of the E-theory does not explain why.

In addition, there are arguably hard problems surrounding the distinction between perceptual experiences and other kinds of experiences. Perceptual experiences have *apparent presentational character*: they first-personally seem to *present* their targets to the subject. The idea is roughly that how the subject's experience unfolds over time introspectively seems to be determined largely by how its targets themselves unfold over time. The targets of other experiences, such as sensory imaginings and episodic memories, do not introspectively seem to be presented to the subject in this sense. Perhaps relatedly, the targets of perceptual experiences seem to be *real, mind-independent particulars* rather than merely imaginary particulars or universals. So consider these questions:

CQ10. Why does your perception of the scarlet pepper have the phenomenology associated with *perception* rather than having the phenomenology associated with sensory imagination?

CQ11. Why does your perception of the scarlet pepper seem to be an experience of a *mind-independent particular* rather than merely an experience of a mind-dependent particular or an experience of universals?

Some will worry that CQ10–CQ11 constitute further hard problems of consciousness, problems whose existence the E-theory does not explain.[13]

I ask you to reserve judgment about these worries. I will offer a complete, reductive answer to CQ9 in chapter 7 and complete, reductive answers to CQ10 and CQ11 in chapter 8. If these accounts are correct, then CQ9–CQ11 do not constitute *new* hard problems; any hard problems here reduce to those constituted by CQ1 and CQ2.

Conclusion There is much to explain about which questions do, and do not, constitute hard problems. The E-theory is the simplest theory we have found that can do this explanatory work. Until we find a better option, we should adopt it.

Now that we have a theory of what generates the hard problems, we can inquire into the consequences of that theory. I begin by considering how the E-theory relates to the central thesis defended in chapters 2 and 3: extended partial revelation.

4.5 The E-Theory and Extended Partial Revelation

In this section, I will argue that accepting the E-theory gives us two additional reasons to accept extended partial revelation. These reasons, although not decisive, are substantial, and they are also independent of the arguments for extended partial revelation found in previous chapters. The first reason will be that accepting the E-theory would give us very strong reason to accept an especially controversial consequence of extended partial revelation. The second reason will be that extended partial revelation would provide answers to some large outstanding questions for the E-theorist. In the final analysis, then, the E-theory and extended partial revelation will make a very attractive package—which, incidentally, will lead me to a final refinement of the E-theory itself.

The first way that the E-theory supports extended partial revelation Extended partial revelation is the claim that, in the first instance, all sensory experiences reveal substantial essential truths about the sensory qualities and nothing else. Because it is a priori that essential truths hold with metaphysical necessity, extended partial revelation entails:

> *Revelation of necessities*: All sensory experiences reveal truths of the form <Necessarily, sensory quality *Q* is *F*>.

Revelation of necessities is obviously a very controversial claim. However, I will argue that we have very strong reason to accept revelation of necessities if we accept the E-theory, and that, as a consequence, the E-theory supports extended partial revelation.

To see why the E-theorist should accept revelation of necessities, recall that the E-theorist accepts the following commitment:

E1*. It strongly initially seems impossible to us for purely physical facts to make it the case that a subject has a conscious experience of one sensory quality rather than another.

Now suppose for the sake of reductio that revelation of necessities is false. For example, suppose that you experience scarlet, and your experience makes it seem to you that scarlet has a particular property, *F*. Suppose also that, in light of your experience, you do not understand how purely physical facts could make it the case that something is *F* rather than *G*. However, *you have no idea whether scarlet is necessarily F*. For all that your experience says, something can be scarlet without being *F*. (*F* might be the property of being Edgar Allan Poe's favorite sensory quality.) Now recall this question about your experience of a scarlet pepper:

CQ2. Why do you experience scarlet rather than experiencing lime green?

CQ2 constitutes a hard problem of consciousness. Does our current theory predict this? No! To be sure, it predicts that you will find it hard to understand why scarlet is *F* rather than *G*. However, our current theory predicts that you will have no particular reason to think that something must be *F* in order to be scarlet. Thus, it does not predict that you will find it hard to understand why you experience scarlet rather than experiencing lime green.

Thus, if the E-theorist is to explain why CQ2 constitutes a hard problem of consciousness, they must accept revelation of necessities.

Another way to appreciate the point is return to this question:

EQ2. Why does the pepper exist rather than not existing?

Experience makes it seem to us that this pepper is scarlet, and we do not understand how purely physical facts could make it the case that something is scarlet rather than lime green. Nevertheless, EQ2 does not constitute a hard problem. How is this possible?

The answer is that it does not seem to us, on the basis of experience, that the pepper is *necessarily* scarlet. That clears away any obstacles in understanding how purely physical facts could make it the case that this pepper exists—we can consider the possibility that the pepper exists *without being scarlet*. Similarly, for CQ2 to constitute a hard problem, it is not enough for experience to make it seem to us that scarlet is *F* and for us to find it hard to understand how purely physical facts could make something *F*. Experience must also make it seem that scarlet is *necessarily* F. Thus, the E-theorist must accept revelation of necessities for their explanations to go through.

You might wonder whether similar considerations support the claim that experience does not reveal necessary truths about anything besides the sensory qualities. The argument would be this. Suppose, for the sake of reductio, that experience *does* reveal necessary truths about other entities—about objects, say. Then a hard problem should be constituted by CQ4—the question, Why do you experience this pepper rather than an intrinsically identical pepper in Vietnam? But CQ4 does not constitute a hard problem. Thus, experience does not reveal necessary truths about objects.

This argument is not sound. Even if someone thinks that experience *does* reveal necessary truths about objects, they have a perfectly respectable way of explaining why CQ4 does not constitute a hard problem: they can say that we have no trouble seeing how purely physical facts could make it the case that those necessary conditions obtain. That is why I stop at the claim that the E-theorist should accept revelation of necessities.

Next, I submit that since revelation of necessities is in turn entailed by extended partial revelation, the E-theory also supports extended partial revelation. To be sure, even if *q* is entailed by *p*, support for *q* is not *always* support for *p*. <There are at least 100 students> entails <there are at least 50 students>, and <there are exactly 75 students> entails the truth of the second claim, but it also entails the *falsity* of the first claim. Still, nothing like that is going on here. It is not as though the E-theory somehow confirms revelation of necessities while disconfirming extended partial revelation more generally. More importantly, revelation of necessities is an especially controversial entailment of extended partial revelation. It is very controversial to say that experience reveals not just how sensory qualities are but how they are *as a matter of metaphysical necessity*. Given these background circumstances, the E-theory's very strong support for revelation of necessities amounts to significant support for extended partial revelation, as well.

The second way that the E-theory supports extended partial revelation Suppose that the E-theory is true. Even so, we are left with some important questions. To begin with, the E-theory says that we find it hard to understand (i) how purely physical facts could ground the existence of a conscious experience and (ii) how purely physical facts could ground an experience of one sensory quality rather than another. But why, exactly, do we find these things hard to understand? In addition, the E-theory says that we do *not* find it hard to understand how purely physical facts could ground an experience of one *object*—or one *property-instance* or one *high-level property*—rather than another. But why don't we find these things hard to understand? What distinguishes experiences of objects, property-instances, and high-level properties, on the one hand, from experiences of sensory qualities, on the other hand?

I will argue that we can answer all of these questions in a satisfying way if we extend the E-theory by accepting extended partial revelation, plus some other claims that are independently plausible in light of extended partial revelation. This gives the E-theorist an additional reason to accept extended partial revelation.

Here are the claims that I think we should accept. This package of views constitutes the final version of the E-theory:

> *Extended partial revelation*: In the first instance, all sensory experiences reveal substantial essential truths about the sensory qualities and nothing else.

E1. Often, it strongly initially seems impossible to us for purely physical entities to satisfy the essential conditions that are revealed to us in sensory experience (as described in extended partial revelation). When and because this happens, there are hard problems.

E2. It strongly initially seems impossible to us for purely physical facts to make it the case that there exists partly revelatory awareness (as described in extended partial revelation). When and because this happens, there are hard problems of consciousness.

E3. There are no other sources of hard problems of consciousness or the external world.

Let me first argue that extended partial revelation makes each of E1–E3 very plausible. This lowers the theoretical cost of accepting these other claims, if we already accept extended partial revelation.

Assuming that extended partial revelation is true, E1 is very plausible. For if extended partial revelation is true, then experience presumably reveals that it is part of the essence of scarlet to be *that way*—the way that we encounter in experience and that is difficult to articulate in nondemonstrative terms. And it is obviously very hard to understand how purely physical facts could make it the case that anything is *that way*.

Assuming that extended partial revelation is true, E2 is also very plausible. The argument for this point is somewhat more extended. Recall that for a mental event *M* to *reveal* a truth *p* about some entity *e* is for the following conditions to be met:

(i) Mental event *M* makes the subject aware of entity *e*.

(ii) For a conceptually sophisticated subject, rational reflection on mental event *M* will, by itself, make it seem that *p* is true.

(iii) The subject's awareness of entity *e*, as described in (i), provides an intelligible basis for its seeming that *p* is true, as described in (ii).

I will grant, for the sake of argument, that we have no particular difficulty in seeing how conditions (i) and (ii) could be met. Perhaps we have some difficulty in seeing how genuine *awareness* could arise from purely physical facts, as per (i), but let us waive that. Then it is easy enough to see how rational reflection on that awareness might make some particular proposition, *p*, seem to be true, as per (ii).

The real difficulty arises with condition (iii). When you see a scarlet pepper, your experience will make various claims about the essence of scarlet seem true to you. This would be easy enough to explain if this happened just as a brute matter of fact. But that is not what happens. Seeing the scarlet pepper gives you a rational basis, something from which you can *extract* truths about the essence of scarlet. It is deeply mysterious how this could be a purely physical matter. Look at a scarlet thing and reflect on the kind of awareness you have, and you will see the problem.

It might help to give a contrasting example (which also appeared in appendix 2B). Suppose that you are sitting in a windowless room looking at a computer screen. Outside the room is a radar that scans the objects passing by and detects the degree to which each object has various gradable properties. You have no idea what properties these are. But as the radar does its work, all of this information is collected and sent to your computer screen. Now, the radar might tell you that it has just detected a series of objects. The first object has property *F*, which essentially consists of 20 percent of property 1 and 83 percent of property 2; the second object has property *G*, which essentially consists of 75 percent of property 1 and 54 percent of property 2; and the third object has property *H*, which essentially consists of 97 percent of property 1 and 3 percent of property 2. By itself, rational reflection on the results of the radar will make it seem to you that it is essential to properties *F*, *G*, and *H* that *F* is more similar to *G* than it is to *H* with respect to property 1. It is no mystery how this is possible, for, in a certain sense, you have no idea what properties 1 and 2 are, and thus you have no idea what properties *F*, *G*, and *H* are.

Something completely different is happening when you see objects of different colors. Such experiences give you an intelligible basis for understanding what colors *are*. *This* is what we find difficult to account for in purely physical terms. Thus, given extended partial revelation, E2 is independently plausible.

Finally, we should not posit additional sources of hard problems without necessity. Thus, E3 is a reasonable default view.

Here, again, is the package of claims that I am recommending:

The *E-theory* (final version):

Extended partial revelation: In the first instance, all sensory experiences reveal substantial essential truths about the sensory qualities and nothing else.

E1. Often, it strongly initially seems impossible to us for purely physical entities to satisfy the essential conditions that are revealed to us in sensory experience (as described in extended partial revelation). When and because this happens, there are hard problems.

E2. It strongly initially seems impossible to us for purely physical facts to make it the case that there exists partly revelatory awareness (as described in extended partial revelation). When and because this happens, there are hard problems of consciousness.

E3. There are no other sources of hard problems of consciousness or the external world.

I will now argue that this final version of the E-theory helps us answer the lingering questions we had about the second version of the E-theory. The second version of the E-theory says that we find it hard to understand how purely physical facts could ground the existence of a conscious experience. Our first question was, Why, exactly, do we find this hard to understand? The answer is that conscious experiences are partly revelatory (extended partial revelation), and we find it hard to understand how purely physical facts could ground the existence of something that is partly revelatory (E2).

The second version of the E-theory also says that we find it hard to understand how purely physical facts could ground an experience of one sensory quality rather than another. Our second question was, Why, exactly, do we find hard this hard to understand? The answer is that conscious experiences reveal some essential conditions on the instantiation of sensory qualities (extended partial revelation), and we often find it hard to understand how purely physical entities could satisfy those conditions (E1).

We had one last question about the second version of the E-theory. The theory says that we do *not* find it hard to understand how purely physical facts could ground an experience of one *object*—or one *property-instance* or one *high-level property*—rather than another. Why don't we find these things hard to understand? What distinguishes experiences of objects, property-instances, and high-level properties, on the one hand, from experiences of sensory qualities, on the other hand? The answer is that experiences do *not* reveal the essences of particulars, property-instances, or high-level properties (extended partial revelation). Thus, there is nothing that would generate hard problems about these entities (E3).

In short, the earlier version of the E-theory leaves us with some large outstanding questions. Extended partial revelation would help us answer these questions. That is an additional reason for the E-theorist to accept extended partial revelation and to accept the finalized version of the E-theory more broadly.

Conclusion An especially controversial consequence of extended partial revelation is revelation of necessities, which says that all sensory experiences reveal truths of the form <Necessarily, sensory quality *Q* is *F*>. If we accept the second version of the E-theory, then we have very strong reason to accept this claim. In addition, the second version of the E-theory states that we find certain things hard to understand, while not finding other, very similar things hard to understand. What generates these asymmetries? We can answer these questions by accepting extended partial revelation, plus some further claims made plausible by extended partial revelation. For both of these reasons, the arguments for the second version of the E-theory provided in this chapter are—to a lesser but still substantial extent—arguments for extended partial revelation. More broadly, they are arguments for the final version of the E-theory.

Notice, by the way, that the E-theory, in its final version, is compatible with either physicalism or anti-physicalism, regarding both the sensory qualities and our experiences of them. The theory only explains why it *strongly initially appears* to be difficult for purely physical facts to fully ground facts about sensory qualities and our experiences of them. This leaves it open whether the difficulties can ultimately be surmounted within a physicalist framework or whether we must reject physicalism itself. I will maintain this neutrality throughout this book, in reflection of my genuine uncertainty.[14]

Looking forward The E-theory tells us what generates the hard problems of consciousness. I want to leverage those results into some conclusions about the very nature of one important form of consciousness.

4.6 Hard Character and Hard Consciousness Defined

In this section, I will define a pair of terms: **hard character**, which is roughly the property of experience that generates the hard problems of consciousness, and **hard consciousness**, which is roughly the variety of consciousness associated with the hard problems of consciousness. My definitions will not be up for dispute: anyone can define terms however they please. What *can* be disputed is whether my definitions are the most productive ones. The thesis of this section is that they are. Unlike various alternatives, my definitions clearly refer, and their referents are clearly of philosophical significance. This sets up the discussion of the next section, where I offer a substantive theory of both hard character and hard consciousness.

Consider some much-examined problems that arise for certain experiences. In particular:

1. The *explanatory gap problem*: We seem unable to understand how the experience could be nothing more than something that is conceived purely physically and/or functionally.
2. The *knowledge problem*: We seem unable to know everything about the experience just on the basis of having all physically and functionally conceived information.
3. The *conceivable absence problem*: We seem to be able to conceive of the experience as present in one subject *S* yet absent in some other subject who is identical to *S* in all functional and physical respects.
4. The *conceivable inversion problem*: We seem to be able to conceive of the experience as invertible across subjects who are identical in all functional and physical respects.[15]

Problems 1–4 seem to be problems about how it is possible to have a certain kind of *awareness* or *consciousness*. Thus, here are some definitions that initially seem to lock onto some important phenomena:

> *First proposal*: The *hard character* of an experience is the property of that experience that generates these problems, and for a mental item to be *hard-conscious* is for it to have hard character.

You might wonder why I am using the terms *hard character/consciousness* instead of *phenomenal character/consciousness*. The short answer is that I believe that different philosophers have unwittingly used the expression "phenomenal character" in extremely different ways, resulting in great confusion. I will defend this claim at length in chapter 8, but in the meantime, there is surely no harm in working with another term.

The first proposal is vague on one point: to have hard character, must a mental item generate *all* of problems 1–4 or just some of them? Here are some more precise proposals:

> *Second proposal*: The *hard character* of a mental item is the property of that mental item that generates *all* of problems 1–4, and for a mental item to be *hard-conscious* is for it to have hard character.
>
> *Third proposal*: The *hard character* of a mental item is the property of that mental item that generates *enough* of problems 1–4, and for a mental item to be *hard-conscious* is for it to have hard character.
>
> *Fourth proposal*: The *hard character* of a mental item is the property of that mental item that generates *any* of problems 1–4, and for a mental item to be *hard-conscious* is for it to have hard character.

None of these proposals is especially productive. We should not say "all of," because some experiences, though they do not generate *all* of the hard problems of consciousness, obviously possess the property that we are trying to latch onto. There is, for instance, no inversion problem associated with olfactory experiences, given the radical asymmetry of the qualitative space associated with smells, but olfactory experiences are paradigms of experiences that generate hard problems. Nor should we say "enough of" or "any of," since some properties that have nothing to do with experience could in principle generate several of problems 1–4. For instance, it is sometimes thought that normative properties, such as moral goodness, generate an explanatory gap problem. Even if that is true, we should not say that if a mental item is morally good, it *thereby* has hard character.

I have a different proposal. Let a *natural kind* be a kind that is, in some important sense, metaphysically real. A natural kind is not gerrymandered or metaphysically arbitrary. Rather, it gets at the structure of reality, carving it at its joints. The properties of having mass, being an electron, and being perfectly circular all plausibly correspond to natural kinds. The property of being either an electron or an abstract object plausibly does not correspond to a natural kind. Relatedly, let a *natural kind term* be a term that is intended to refer to a natural kind—for example, the terms *water* or *gold*.[16]

I propose to define *hard character* as a natural kind term. I would start by identifying the *apparent paradigms* of hard characters as those entities that generate enough of, but not necessarily all of, problems 1–4. I would then use those apparent paradigms to lock onto the relevant natural kind. This solves the problems. It lets us pull in hard characters even if they do not generate any of problems 1–4, as long as those properties belong to the right natural kind. It also lets us push out properties that generate enough of problems 1–4—normative properties, say—if they do not belong to the right natural kind.

More precisely, my proposal is this:

> *Final proposal (part 1)*: If there is a single natural kind property of experience that tends to generate enough of problems 1–4 in paradigmatic cases, then the *hard character* of a mental item is the item's property that belongs to this natural kind.

It is logically possible that there are no natural kinds, or many, that play enough of roles 1–4.[17] To accommodate these possibilities, I would add:

> *Final proposal (part 2)*: If there is no natural kind that tends to generate enough of problems 1–4 in paradigmatic cases, then *hard character* simply refers to any property of that experience that generates enough of these problems. If there several natural kinds that tend to generate enough of problems 1–4 in paradigmatic cases, then *hard character* refers indeterminately to these.

I would also retain the definition of hard consciousness:

> *Final proposal (part 3)*: For a mental item to be *hard-conscious* is for it to have hard character.

I can now outline my case that this definition of the term *hard character* is the most productive one. (Precisely the same case could be made for the term *hard consciousness*.) First, I would assert that hard character, as I have defined it, exists. After all, it should be extremely plausible that problems 1–4 arise *sometimes*: earlier in this chapter, we have examined many particular hard problems of consciousness. Next, I would observe that given that problems 1–4 arise at all, the most parsimonious view is that they are generated by a single natural kind property. Indeed, in the next section, I will argue in detail that this parsimonious view is correct. Given this conclusion, we can say that not only does the term *hard character* refer, but it refers to a natural kind property—a property that is metaphysically real, carving reality at its joints.

However, you might worry that, somewhere in my reasoning, I have made a false presupposition. Let us consider several objections in this vein.[18]

Objection I say that hard character exists. But this presupposes that there is some property of my experience that *in fact* generates explanatory gaps, missing knowledge, and the like. This presupposition is very controversial.

Reply I do not make this presupposition. By saying that experiences have hard character, I assume only that there is a property of experience that has *seemed—in a reasonably robust way, to a reasonably large group of philosophers, from the late twentieth century to the early twenty-first century*—to generate explanatory gaps, missing knowledge, and the like. (Let the phrase between the dashes be a specification of the meaning of the word "we" in problems 1–4.) This assumption is obviously true.

Objection Qualia theorists have usually assumed that problems 1–4 are wholly generated by intrinsic properties of the subject, such as neural properties. Meanwhile, many representationalists have assumed that the hard problems are generated by those intrinsic properties *together with* historical facts. But it was once widely presupposed that the hard problems are not generated by any of the subject's *current* relations to external properties and objects (e.g., their current perceptual relation to the pepper and its instance of scarlet). Many naïve realists would dispute this presupposition.[19]

Reply Others may have made this presupposition; I have not. My statements of the four problems, and my definitions, leave open the possibility that problems 1–4 can be generated by the property of standing in a current perceptual relation to the pepper.

Objection It is commonly presupposed that the hard problems are, in the first instance, problems about *experiences* rather than the *targets* of those experiences. The naïve realist might reject this presupposition. She might think that sensory qualities—colors, smells, tastes, and the like—are what generate the hard problems in the first place, with experiences merely inheriting these problems.[20]

Reply To this I would say two things. First, there is nothing in my definition of hard character that presupposes that experiences *are* the initial source of the hard problems. Suppose that sensory qualities are the initial source. Still, there would remain the hard problem of explaining how it is possible to have an experience of scarlet in a purely physical world, and the term *hard character* would still refer. Even the naïve realist should concede this: they might say that one hard character is the property—had by the experience—of being a primitive perceptual relation to an instance of scarlet.

My second reply is less concessive: I have already argued that *some* hard problems have nothing to do with the sensory qualities. I have argued that even if we understood exactly how a purely physical pepper could be scarlet, there would remain a hard problem about how we could *experience* scarlet. I have also argued that even though we understand perfectly well how a purely physical object can be square, there remains a hard problem about how subjects can *experience* squareness.

I conclude that there are no false presuppositions in my descriptions of problems 1–4 or in my definitions of hard character and hard consciousness.

Conclusion In this section, I have introduced a pair of definitions. First, a definition of hard character:

> If there is a single natural kind property of experience that tends to generate enough of problems 1–4 in paradigmatic cases, then the *hard character* of a mental item is the item's property that belongs to this natural kind. If there is no natural kind that generates enough of problems 1–4, then *hard character* simply refers to any property of that experience that generates enough of these problems. If there are several natural kinds that generate enough of problems 1–4, then *hard character* refers indeterminately to these.

Second, a definition of hard consciousness:

> For a mental item to be *hard-conscious* is for it to have hard character.

Definitions are cheap. What is valuable is the claim that these definitions are *productive*. I have defended this claim carefully, partly by showing that these definitions yield the intuitively correct extensions for the terms and partly by rebutting a series of arguments against the very existence of hard character/hard consciousness, as defined here.

In previous chapters, I have mentioned consciousness several times. I can now clarify that I was uniformly talking about *hard* consciousness. I will identify other forms of consciousness in later chapters.

To define a term is not to give a substantive theory of its referent. Thus, the next obvious question is, What are hard character and hard consciousness, substantively speaking?

4.7 A Theory of Hard Character and Hard Consciousness

The purpose of this section is to defend this answer:

> For a mental item to have *hard character*—equivalently, for it to be *hard-conscious*—is for it to be at least partly constituted by partly essence-revealing awareness.

Notice that this theory maintains neutrality about several important questions. First, it is neutral about whether sensory experiences are the only hard-conscious mental items or whether there are further mental items, such as thoughts, that are also hard-conscious. Those who think that thoughts can be hard-conscious might say that *in sensory experience*, the only targets of partly essence-revealing awareness are sensory qualities, while *in thought*, there is partly essence-revealing awareness of other targets, such as meanings. Second, my theory of hard character is neutral between **hard-character internalism**, which says that hard character supervenes on the intrinsic properties of the subject, and **hard-character externalism**, which denies this. A hard-character internalist can accept my theory by saying that subjects who are intrinsically identical are necessarily identical in what kinds of partly essence-revealing awareness they enjoy.

My argument for this theory of hard character will be this ("PRA" stands for "partly revelatory awareness"):

PRA1. For a mental item to have hard character—equivalently, for it to be hard-conscious—is for it to have the natural kind property (assuming there is one) that tends to generate enough of problems 1–4 in paradigmatic cases.

PRA2. The property of being at least partly constituted by partly essence-revealing awareness is a natural kind property.

PRA3. The property of being at least partly constituted by partly essence-revealing awareness is the property that generates the hard problems of consciousness (in my technical sense).

PRA4. If something generates the hard problems of consciousness (in my technical sense), then it tends to generate problems 1–4 in paradigmatic cases.

PRA5. For a mental item to have hard character—equivalently, for it to be hard-conscious—is for it to be at least partly constituted by partly essence-revealing awareness.

Let us first examine the premises of this argument and then the inference (which is not quite deductively valid).

According to PRA1, for something to have hard character—equivalently, for it to be hard-conscious—is for it to have the natural kind property (assuming there is one) that tends to generate enough of problems 1–4 in paradigmatic cases. This follows immediately from the definitions of hard character and hard consciousness defended in the previous section.

According to PRA2, the property of being at least partly constituted by partly essence-revealing awareness is a natural kind property—that is, a property that is metaphysically real; one that carves reality at the joints, rather than at arbitrary points. (There is no implicature that a natural kind property is natural in the sense of being *physical*. Dualists can happily accept PRA2.) This premise is obviously true: the property of being at least partly constituted by partly essence-revealing awareness is not arbitrary but deeply metaphysically significant.

The next premise of the argument, PRA3, says that the property of being at least partly constituted by partly essence-revealing awareness is the property that generates the hard problems of consciousness, in my technical sense. In defense of this premise, recall our theory, from section 4.5, of what generates the hard problems of consciousness:

The *E-theory* (final version):

Extended partial revelation: In the first instance, all sensory experiences reveal substantial essential truths about the sensory qualities and nothing else.

E1. Often, it strongly initially seems impossible to us for purely physical entities to satisfy the essential conditions that are revealed to us in sensory experience (as described in extended partial revelation). When and because this happens, there are hard problems.

E2. It strongly initially seems impossible to us for purely physical facts to make it the case that there exists partly revelatory awareness (as described in extended partial revelation). When and because this happens, there are hard problems of consciousness.

E3. There are no other sources of hard problems of consciousness or the external world.

Notice that E1 can be triggered only if and because a mental item is partly constituted by partly essence-revealing awareness—otherwise, no essential conditions will be revealed to us in the first place. Likewise, E2 can be triggered only if and because a mental item is partly constituted by partly essence-revealing awareness—otherwise, there can be no question of how such awareness exists. Thus, the E-theory entails that a mental item will generate a hard problem of consciousness only if, and because, it is partly constituted by partly essence-revealing awareness.

The final premise of the argument, PRA4, says that if something generates the hard problems of consciousness (in my technical sense), then it tends to generate problems 1–4 in paradigmatic cases. To see why we should accept this premise, recall that for there to be a **hard problem** (in my technical sense) is for the following to be true, where p, q, and r range over propositions:

(i) we have strong background evidence that *p* fully grounds *q*—but, at the same time,

(ii) it strongly initially seems impossible to us that *p* fully grounds *q*—we cannot see why *p* would make it the case that *q* rather than making it the case that *r*.

Now suppose for the sake of argument that there is a hard problem, in my sense. Suppose, for example, that we cannot see how purely physical facts would make it the case that you are having an experience (which is by definition hard-conscious) of scarlet, rather than having an experience of lime green or merely unconscious information processing. First, this would by itself constitute an *explanatory gap* between (a) the complete physical account of your experience, and of the pepper, and (b) the way that your experience seems to be from the inside. In particular, it would strongly initially seem to us that (a) did not fully ground (b). Second, knowing all of the physical facts about your experience, and about the pepper, would seem insufficient for knowing *everything* about your experience: we would be mystified about why you were having a conscious experience and why you were experiencing scarlet rather than lime green. Third and fourth, we would seem to be able to conceive, in arbitrary detail, of a subject who is functionally and physically just like you when you experience the scarlet pepper but who did not have any kind of experience whatsoever, and as long as the relevant quality space is symmetrical along some axis (e.g., red–green), we will be able to conceive of a subject who is functionally and physically just like you but who experiences lime green rather than scarlet. In other words, we would face conceivable absence and conceivable inversion problems.

Thus, generating a hard problem, in my technical sense, will suffice to generate problems 1–4 in paradigmatic cases.

That completes my defense of PRA1–PRA4. This does not quite entail the conclusion, PRA5, which says that for a mental item to have hard character—equivalently, for it to be hard-conscious—is for it to be at least partly constituted by partly essence-revealing awareness. For PRA1–PRA4 entail only that there is *at least one* natural kind property that generates problems 1–4 in paradigmatic cases. Unlike PRA5, they do not entail that this is the *only* natural kind property of this sort. Still, this is the most parsimonious view: what reason could there be to posit further natural kind properties of this sort? Thus, we have sufficient reason to accept PRA5.

Conclusion, and a range of applications By definition, if there is a single natural kind property that tends to generate enough of problems 1–4 in paradigmatic cases, then hard character is that property (PRA1). I have argued that what fits the bill is the property of being at least partly constituted by partly essence-revealing awareness. For this is a natural kind property (PRA2); it is what generates the hard problems of consciousness, in my technical sense (PRA3), and generating those problems suffices for generating problems 1–4 (PRA4). As a corollary, I note that the particular hard character of an experience will be its property of the form, *being partly constituted by partly essence-revealing awareness of* ___.

I observe in passing that this has an important application to (the many and large) debates about what, in particular, constitutes the hard characters of various experiences. For example, when I see a pine tree, is there any hard character over and above that associated with my seeing various colors, shapes, and textures—is there some further hard character associated with my seeing the pine tree *as a pine tree*? The current theory says no: hard character is constituted by partly essence-revealing awareness, and as I argued in chapter 3, seeing a pine tree does not reveal any part of the essence of the property of being a pine tree. I leave it to others to apply the theory to other debates—for instance, to the debate over whether there is any hard character associated with cognition, over and above the hard character associated with sensory experience.

4.8 Some Objections to the Significance of Hard Character

I have said that hard character not only exists but is a natural kind. Some naïve realists disagree: they argue that even if hard character exists, it is of limited theoretical interest. In this section, I rebut two very similar naïve realist objections to this effect.

First objection This objection begins with this premise ("AHC" stands for "against hard character"):

AHC1. Conscious perception makes us sensorily aware of external objects and properties, not the perception itself.

AHC1 is intended to be immediately plausible. From this, my opponent infers:

AHC2. It is a mistake to characterize this sensory awareness in terms of any property of the perception, such as hard character.

For given AHC1, my opponent thinks that we should surely characterize sensory awareness by appealing directly to the external objects and properties themselves, rather than to a property of the perception. Finally, from AHC2, my opponent draws their conclusion:

AHC3. Any property of the perception, such as hard character, is of limited theoretical interest.

For surely the point of hard character was to characterize the kind of sensory awareness involved in conscious perception.[21]

I accept AHC1—again, the claim that conscious perception makes us sensorily aware of external objects and properties, not the perception itself. Does this support AHC2, the claim that we should characterize this sensory awareness *purely* in terms of external objects and properties, without appealing *at all* to properties of the perception? Clearly not: a scarlet pepper might be right in front of me, but if my eyes are closed, I will have no sensory awareness of the pepper or its properties. So properties of my perception must figure *somehow* in a perspicuous characterization of my sensory awareness.

At most, what AHC1 supports is the conclusion that sensory awareness should be *at least partly* characterized in terms of external objects and properties. That is compatible with *also* partly characterizing sensory awareness in terms of properties of the perception, such as hard character. That is just what our current theory does. It says that the hard character of your conscious perception of a scarlet pepper is the property of *being partly constituted by partly revelatory awareness of scarlet, pepper-shapedness, and. . . .* This property concerns external properties like scarlet and pepper-shapedness.

Still, my opponent might observe that *external objects like peppers* do not figure in hard character. Yet surely we must appeal to such objects to fully characterize perceptions of objects. Thus, they might say, hard character remains of limited theoretical interest.

I agree that we must appeal to external objects to fully characterize perceptions of objects. But I deny that this undermines the theoretical interest of hard character. The proper conclusion is that hard character does not, *by itself,* fully characterize perceptions of objects. But so what? To be of theoretical interest, hard character need only *partly* characterize perceptions of objects. In later chapters, I will explain where I think external objects enter the picture.

Second objection Some have developed a similar objection that turns on the *transparency of experience*. Roughly speaking, to say that experience is *transparent* is to say that when we introspect on our experiences, we cannot directly attend to any properties of the experience itself; we can only directly attend to the external objects and properties that we are experiencing. With this in mind, my opponent asserts:

AHC1*. The transparency of a conscious perception is explained by the actual presence of external objects and properties.

This premise is supposed to be immediately plausible. From there, my opponent observes that the actual presence of external objects and properties is not a property of the perception. Thus, they arrive at:

AHC2*. The transparency of a conscious perception is not explained by any property of the perception. (Supported by AHC1*.)

And since we might have regarded the transparency of perception as an especially central explanatory target, my opponent draws this conclusion:

AHC3*. Any property of the perception, such as hard character, is of limited theoretical interest. (Supported by AHC2*.)[22]

My response to this objection mirrors my response above. I will grant for the sake of discussion that the transparency of a conscious perception is explained by the actual presence of external objects and properties, as per AHC1*. I find it unwise to infer from this that the transparency of a conscious perception is not explained by any property of the perception, as per AHC2*. External objects and properties can be present even when I am entirely unconscious and thus free of any experiences. Thus, properties of my perception must be at least *partly* responsible for the transparency of my experience. At most, what AHC1* suggests is that the transparency of experience should be explained by a property of my perception that involves a *relation* between it and external entities.

Hard character is one such property. I emphasize again that it cannot be the *only* such property. Hard character concerns only the sensory qualities, but the transparency of experience also pertains to external particulars. But, again, this point does not undermine the theoretical significance of hard character. It just shows that the transparency of experience cannot be *exhaustively* explained by hard character. Hard character can still be an important part of the story.

Conclusion Hard character is a property of hard-conscious perceptions (and other experiences), and these experiences make us sensorily aware of external particulars and properties. But this sensory awareness cannot be explained *purely* by the presence of those external things: much of what is present in my environment goes unperceived. Properties of perceptions must be part of the explanation, too. Thus, hard character can retain its theoretical significance.

To be sure, hard character is not at all constituted by external particulars and so cannot explain why hard-conscious perceptions make us sensorily aware of these. That is not a problem: it just means that hard character does not, by itself, fully characterize hard-conscious perceptions. I will fill in the rest of the characterization in later chapters.

4.9 Conclusion

In this chapter, I have had three entwined purposes.

My first purpose has been to develop a theory of what generates the hard problems of consciousness and the external world. In service of this purpose, I developed the E-theory. This theory says that in the first instance, all sensory experiences provide partly revelatory awareness of sensory qualities and of nothing else (extended partial revelation). This generates hard problems in two ways. First, in light of what is revealed to us about the essences of the sensory qualities, we find it mysterious how instantiations of any sensory qualities could be fully grounded in the physical facts (E1). Second, we find it mysterious how the existence of partly revelatory awareness itself could be fully grounded in the physical facts (E2). The hard problems of the external world are generated purely by the first source, while the hard problems of consciousness may be generated by either source; there are no other sources of hard problems of consciousness or the external world (E3).

My second purpose has been to provide multiple new lines of support for extended partial revelation. In service of this purpose, I argued that extended partial revelation almost immediately delivers an explanation of which questions constitute hard problems of the external world. I also argued that the E-theory strongly supports an especially controversial consequence of extended partial revelation and that extended partial revelation helps us answer some pressing questions about the E-theory. These are powerful new reasons to accept extended partial revelation.

My third purpose has been to identify the most productive definitions of the terms *hard character* and *hard consciousness*, and to provide a substantive account of their referents. In service of this purpose, my basic strategy was to define *hard character* as, roughly, the natural kind property that generates the hard problems of consciousness and *hard consciousness* as the kind of consciousness corresponding to hard character. I argued that, so defined, these terms refer, and their referents are philosophically significant. I went on to defend a substantive theory of these properties: what it is for a mental item to have hard character, or to be hard-conscious, is for it to be at least partly constituted by partly essence-revealing awareness.

This concludes part I of this book. A central purpose of this part has been to defend extended partial revelation. I have also identified the theories that can explain this thesis: impure revelatory monism, which comes in representationalist and naïve realist varieties; pluralistic representationalism; pluralistic naïve realism; strange pluralism; and rich pluralism.

It is the last of these theories that I prefer, and in parts II and III, I explain why.

II Rich Pluralism versus Representationalism

As I define things, *representationalism* is the view that (i) to perceive is to deploy at least one kind of awareness, typically of external particulars and their properties/property-instances, and (ii) every kind of awareness constitutive of perception is representational (i.e., subject to a particular kind of error). In this second part of the book—which is very short, consisting only of chapter 5—I will make a detailed comparison between various versions of representationalism and rich pluralism. (I mostly leave comparisons with naïve realism to part III.)

On the one side, I will argue that there is little favoring representationalism over rich pluralism. The rich pluralist can easily honor all of the core motivations for representationalism, and—surprisingly—adequate representationalist theories make no gains in parsimony over rich pluralism. On the other side, I will argue that the best representationalist theories have a major drawback: they predict that experience is subject to a particular form of error that we do not actually find. Rich pluralism makes the correct prediction.

My conclusion will be that rich pluralism is superior to any representationalist theory of perception.

5 Perception, Hallucination, and Illusion

In previous chapters, I used data about what is revealed to us in experience, and about what hard problems there are, to defend extended partial revelation. This winnows out a wide range of theories of perception, but it leaves us with plenty of theoretical options. Which of these should we choose? In this chapter, I compare rich pluralism to the best remaining *representationalist* options: impure revelatory representationalism and pluralistic representationalism. By way of preparation, I provide a new definition of the paired notions of representations and contents (section 5.1), flesh out the rich pluralist's theory of sensory representation (section 5.2), and use this to give an account of the natures of perception (section 5.3), hallucination (section 5.4), and—briefly—illusion (section 5.5). Then, after a brief intermission (section 5.6), I begin my assessment.

I contend that impure revelatory representationalism and pluralistic representationalism have no particular advantage over rich pluralism. The main motivation for representationalism is that it can give a satisfying account of hallucination. However, rich pluralists accept the core posits of representationalism—they just introduce additional posits about deep awareness—so they enjoy these same explanatory benefits (section 5.7). In addition, not only monistic theories, but also pluralistic ones, can explain the unity of experience (section 5.8). Finally, and perhaps most surprisingly, adequate versions of representationalism are no more parsimonious than rich pluralism. This is obvious for pluralistic representationalism. What is less obvious is that monistic representationalists must complicate their theories to account for hard-*unconscious* perceptions (section 5.9).

However, I argue that impure revelatory representationalism and pluralistic representationalism both come with a large cost: they predict that experiences can err about the very essences of the sensory qualities—but this is incorrect. Rich pluralism gets things right (section 5.10), while also explaining why experience sometimes errs about whether sensory qualities are instantiated (section 5.11).

On the basis of this holistic assessment, I conclude that rich pluralism is superior to any representationalist theory of perception (section 5.12).

5.1 Representations and Contents

At a minimum, a **representation** is an entity that is *about* something—its *content*. But what more can we say to define the paired notions of representations and contents?

For my purposes in this book, it will be useful to define these terms in such a way that all representationalists will accept, and all naïve realists will deny, that conscious perceptions are representations—equivalently, that conscious perceptions have contents. However, some familiar definitions do not permit this.

For instance, one standard option is the *identity definition*. The idea is, first, to define *representationalism* about some experience *E* as the theory that having *E* is identical to standing in a relation, *R*, which is distinct from and more basic than belief, to a proposition, *p*. We then stipulate that an experience has a *content* (namely, *p*) and is thus a *representation*, just in case this theory is true of it.[1]

The identity definition does not serve my purposes: it allows some naïve realists to accept that conscious perceptions are representations. Consider, for example, a naïve realist who holds (i) that what it is to have a particular conscious perception is to stand in the relation of acquaintance, which is distinct from and more basic than belief, to a certain fact, and (ii) that facts are identical to true propositions. This view—call it *factual naïve realism*—entails representationalism, as characterized by the identity definition.

A second standard option is the *accuracy definition*, which defines representations as having two features. First, their contents are conveyed to the subject. Second, their contents are accurate or inaccurate (or simply true or false).[2] Note that this second condition is satisfied even if a given type of representation has contents that are always accurate: an accurate content is trivially either accurate or inaccurate.[3]

Like the identity definition, the accuracy definition tells us that factual naïve realists are representationalists. After all, the factual naïve realist holds that conscious perceptions convey facts to their subjects, and these facts are, in a trivial sense, always accurate.[4]

I conclude that the identity definition and the accuracy definition may be useful for some purposes. But they are not useful for mine.

To construct a definition that suits my purposes, we should ask, What really distinguishes representationalism from naïve realism? I think it is the idea that conscious perception is *subject to error*. Representationalists and naïve realists agree that (veridical) conscious perceptions are not *in fact* in error about anything. But representationalists think that conscious perceptions *could* have been in error: that is what happens in conscious hallucinations. Naïve realists disagree. They think that conscious hallucinations are not the error-ridden versions of conscious perceptions; conscious perceptions do not have error-ridden versions.

I therefore propose the *error definition*: I understand *representations* as contentful mental states or events that are *subject to error*—they are entities of a kind that can go right or wrong. The claim is not that every *particular* content of a representation might be true or false: the content that $3+5=8$ is necessarily true.[5] The claim is rather that for any representation of a natural, broad kind (such as belief), there are metaphysically possible representations of that kind with true contents (e.g., a belief that $2+2=4$) and metaphysically possible representations of that same kind with false contents (e.g., a belief that $2+2=5$). Incidentally, I do not assume that contents must be metaphysically possible: perhaps it is possible to represent that there are square circles.

The error definition lets us say, at last, that all representationalists affirm that conscious perceptions are representations and that all naïve realists deny this. For representationalists think that conscious perceptions and conscious hallucinations are representations of a single natural, broad kind, *experiencing*. Conscious perceptions are error-free representations of this kind; conscious hallucinations are (one type of) error-ridden representation of this kind. Naïve realists deny this.

The error definition has a further virtue: the demarcation it makes is *natural*—equivalently, it is structural, metaphysically real, joint-carving. For the distinction between mental states or events that are subject to this particular kind of error, and mental states or events that are not, is not arbitrary. It is metaphysically real.

With this definitional point settled, I can begin to flesh out the rich pluralist's theory.

5.2 Sensory Representations

We undergo **sensory events** of many kinds, including hallucinations, sensations, sensory imaginings, genuine episodic memories, and merely apparent episodic "memories." Arguably, each kind of sensory event comes in hard-conscious and hard-unconscious varieties.

The rich pluralist holds that all sensory events are constituted, at least partly, by a particular form of representation, **sensory representation.** Their basic motivation is to respect the very plausible idea that there is a difference in format between (i) sensory events and (ii) nonsensory states or events, such as thoughts, intentions, and desires. Representationalists have offered various accounts of what is distinctive about sensory representation.[6] For the most part, the rich pluralist remains neutral among these.

But they take a stand on one issue, which pertains to the distinction between singular and existential contents. For a content to be *singular* is for it to at least putatively be constituted by a particular, such as an object, a property-instance, or an event. Notice that I am using the expression **singular content** in a broader way than usual: it refers even to contents that are only *putatively* singular, like the content <Santa Claus is jolly>. **Successful singular contents** are actually constituted by one or more particulars—for example, the content <Socrates is a philosopher>, which is constituted by Socrates himself. A representation ends up with a *failed* singular content when it is directed at a particular without actually latching onto one, as with a failed demonstrative term. By contrast with all of this, for a content to be *existential* is for it to not even be directed at a particular. Representations can then be singular or existential in a secondary sense, in virtue of having singular or existential contents.

The rich pluralist says that sensory representations are often singular. When I see a scarlet pepper, I am deploying a sensory representation with a *successful* singular content, a content that is constituted in part by that very pepper and its instances of scarlet and pepper-shapedness. If I hallucinate a scarlet pepper, I am deploying a sensory representation with a *failed* singular content.

The rich pluralist assigns singular contents to typical perceptions, illusions, and hallucinations for the usual reasons. Imagine that I see a blue ball on my left. The ball is illuminated with trick lighting, so I take it to be yellow. I see it through a slanted mirror, so I take it to be directly in front of me. And the mirror is a carnival mirror, so I take the ball to be a cube. In short, on the basis of my perception, I take there to be a yellow cube directly in front of me. In fact, there *is* a yellow cube in front of me, hidden behind the mirror! Despite this, I am having an illusion, not a veridical perception. A simple explanation for this is that my experience has singular content: it represents *the blue ball on my left* as a yellow cube straight ahead.[7]

I incline toward the view that sensory representations are *always* singular. This would permit a more unified account of sensory representation in general: its targets would always include particulars. One challenge to this claim comes from sensory imagination: can't I imagine a gorilla without imagining any gorilla in particular? Of course, but this might be explained in terms of my deploying a failed singular representation, one that does not refer to any gorilla in particular. This does not entail that sensory imaginations are *themselves* failures in any interesting sense: sensory imaginations need not have the function of making successful singular reference. Another challenge comes from nonvisual perception: when I smell the scent of vanilla, are my sensory representations truly *singular*? I find it plausible that they are. They might not sensorily represent any *objects* as being vanilla-scented, but arguably, they still represent particular *instances* of vanilla-scentedness. Still, for official purposes, I will remain neutral about whether sensory representations are always singular.

In sum, the rich pluralist says that all sensory events—perceptions, hallucinations, sensations, sensory imaginings, genuine episodic memories, and failed episodic "memories"—are at least partly constituted by sensory representations. This is a distinctive form of representation that is often, and perhaps always, singular.

5.3 Perception

Turn now to what distinguishes perception, *whether hard-conscious or hard-unconscious*, from sensory representations of other kinds.

To begin with, I think that perception is not just representational. It also requires a genuine relation to its targets, where a *relation* between two entities requires the existence of both entities. More precisely, I think that it is the representational *vehicle* that must stand in a relation to the targets. In virtue of that, the vehicle has a genuinely singular content: without the relation to the target, that target would not be a constituent of the vehicle's content.

This cannot be the only feature distinguishing perception from other sensory events. Episodic memories, understood factively, are equally relational: their representational vehicles, too, must represent their targets by standing in some kind of relation to them. What I propose, then, is that there is a special **perceptual relation** that distinguishes perception from episodic memory. This yields the following account of the nature of perception:

> To be a **perception** is to be a sensory representation that represents by standing in a perceptual relation to certain targets.

But what is a *perceptual relation*? I suggest that it is a relation between a subject's sensory representations and the target(s) such that (i) the representations represent the target(s) directly, (ii) the representations update in real time, and (iii) the contents of the representations vary with the target's spatial position relative to some part of the subject. (Why does condition (iii) refer to *some part of* the subject? To handle cases of scattered subjects: imagine someone whose body, on Earth, has a quantum link to a pair of eyes on Venus.)

What is the significance of *these* three features in particular? It is that when they are taken together, they have a special kind of explanatory value. To appreciate this, suppose that I see a falling leaf. Now imagine a similar mental event, but subtract one of the defining characteristics of the perceptual relation. Perhaps the mental event does not involve *direct* awareness of its target, contrary to (i): perhaps I am seeing the shadow of the leaf against the tree, without seeing the leaf itself. Or perhaps the event does not update in real time, contrary to (ii): it is more like a frozen image of a leaf mid-fall. In both cases, I submit that it is obvious why it is worth distinguishing these mental events from my perception of the falling leaf.

Or imagine an alien being who has some kind of direct awareness of the leaf "from nowhere": the alien is sensorily aware of the spatial layout of the leaf but lacks any sense of where the leaf is relative to itself. This event is much like my perception of the leaf, except that its content does not vary as the target moves around the subject.[8]

This event arguably belongs to a natural kind in its own right. But it is plausible that there is also a natural kind that includes perceptions while excluding alien mental events such as this. Perception has a distinctively *perspectival* character: it reveals, or purports to reveal, the location of its targets in egocentric space. These targets might be ordinary objects: when I see a black Labrador Retriever, my visual perception reveals where it is *relative to me*. Or the targets might be something else. When I breathe in the scent from a bottle of vanilla, it is not the bottle itself but the *scent* that evidently fills my nose; when I hear a cup smash against the floor, it is the *sound* that is evidently coming from my left. It is because of this perspectival character that visual perception literally provides a *point of view* on its targets. The alien experience does not do that. The difference is surely worth marking.

We can bring the point into relief by observing that all three of the distinctive features of the perceptual relation have a special connection to *action*. If I am not directly aware of the leaf, or if I am not aware of it in real time, or if I am not aware of its position relative to me, then I cannot act on the leaf—by trying to catch it, say—*in the fluid and direct way made possible by perception*. This is not what *makes* perception a natural kind: even for a subject who cannot act, perhaps because they are paralyzed, it would be worth marking the difference between perception and various nearby mental events. But the link to action is a *mark* that perception is a natural kind.

You might wonder whether all of these features are necessary for perception. Consider subjects with *akinetopsia*—motion blindness—who see the world like a series of photographs. These subjects are genuinely perceiving, but you might worry that their experiences do not satisfy condition (ii): the underlying representations do not update in real time. To this I respond that they *do* update in real time, albeit much less frequently than normal perceptions.

Or consider condition (iii), which says that the contents of the representation must vary with the target's spatial position relative to some part of the subject. You might wonder how this condition applies to perceptions of bodily sensations—the feeling of pain in the elbow, say. My answer is that when the subject's elbow is near the rest of their body, the pain will also seem to be near the rest of their body. As they move their elbow, the location of the pain relative to the rest of their body will also seem to change.

The present account of perception has another pair of important virtues, which relates to the well-known distinction between access consciousness and hard consciousness. For a mental state or event to be *access conscious* is for it to be directly available to the subject—for example, for verbal report and action. For a mental state or event to be *hard-conscious* is for it to have the kind of consciousness distinctively associated with the hard problems of consciousness.[9]

It is clear that perceptions can be access-unconscious. This is true of certain perceptions in the dorsal stream that directly guide action without being accessible to the subject.[10] It is also clear that perceptions can be hard-unconscious. Again, dorsal-stream perceptions are instances of this. So are the perceptions of very simple creatures and super-blindseers.[11]

The present account has the virtues of explaining both possibilities. Suppose that there is a sensory representation that represents by standing in a perceptual relation to certain targets. This representation need not be available to the subject for purposes of verbal report and action; it might be available only to certain subpersonal systems, such as motor-control systems. Then the subject would have an access-unconscious perception, as in dorsal-stream perceptions. Or suppose that such a sensory representation does not involve any deep awareness, as in dorsal-stream perceptions and the perceptions of simple or super-blindsighted creatures. Then there would be a hard-unconscious perception (which might or might not be access-conscious).[12]

Incidentally, some will wonder why the rich pluralist accepts the broad view that we sensorily represent property-instances. Why not accept the narrow view that we sensorily represent objects (but not property-instances), while being deeply aware of universals? Here we find one good reason: it is possible to have hard-unconscious perceptions of property-instances. With the broad view in hand, the rich pluralist can say that this is a matter of sensorily representing property-instances. If they accepted the narrow view, it is not clear what explanation they could give. They could not say that having a hard-unconscious perception of property-instances was a matter of being *deeply aware* of property-instances, because deep awareness makes any mental item hard-*conscious*.

Conclusion In this section, I have laid out the rich pluralist's account and identified some of its initial virtues.

According to the rich pluralist, what it is to be a perception is to be a sensory representation that represents by standing in a perceptual relation to certain targets. Here, a *perceptual relation* is a relation between a subject's sensory representations and the target(s) such that (i) the representations represent the target(s) directly, (ii) the representations update in real time, and (iii) the contents of the representations vary with the target's spatial position relative to some part of the subject.

This account has several initial virtues. It is extensionally adequate, it makes it plain why perception is a natural kind and why perception has a special connection to action, and it explains why perceptions can be hard-unconscious, as well as access-unconscious.

Still, this is far from a full defense of the account. That will come later.

5.4 Hallucination

In this section, I turn to hallucination. Again, I will lay out the rich pluralist's account and identify some of its initial virtues.

A familiar representationalist idea is that hallucinations are the failed counterparts of perceptions. How might the rich pluralist develop this idea? An obvious thought is that any hallucination is a sensory representation that is *not* a perception. However, this cannot be the whole story: this condition is met not only by every hallucination but also by every sensory imagining, episodic memory, and sensation. What separates hallucinations from these other kinds of sensory representations?

I accept the orthodox representationalist idea that a sensory representation will qualify as a hallucination only if it plays the same core functional roles as a perception—for instance, the subject must be disposed to act as if the content of the representation obtained. To express this idea more precisely, let us say that for a subject to *take* a mental item to be of a certain mental kind is for that mental item to play, in that subject, the core functional roles that characterize that mental kind. Then we can put the rich pluralist's account like this:

> To be a *hallucination* is to be a sensory representation that is *mistakenly* taken to be a perception.

This account has a number of virtues.

To begin with, there is an intuitive sense in which certain hallucinations *match* certain perceptions. Our account lets us identify what it is for this to occur: it is for the hallucination and the perception to be constituted by deep awareness of the same sensory qualities (if they are hard-conscious), together with sensory representations of the same type, modulo whether these representations are successful (having genuinely singular contents) or failed (having merely putatively singular contents).

In addition, it is obvious that, given a particular mental background, any perception and any matching hallucination would cause the subject to form similar beliefs and take similar actions. Our account lets us explain why: it is because, in general, successful and failed representations of the same type will have similar mental effects.

Another virtue of the account is this. Just as *perceptions* can be hard-conscious or hard-unconscious, so too can **hallucinations** be hard-conscious or hard-unconscious. Our account makes room for both kinds of hallucinations. After all, the rich pluralist says that hard consciousness is constituted by deep (i.e., partly essence-revealing) awareness and that hallucinations—sensory representations that are mistakenly taken as perceptions—can occur with or without deep awareness.

The account has a further virtue. It is a datum that simple creatures can hallucinate. Some views seem to struggle to explain this. For instance, according to the **indiscriminability view**, what it is to hallucinate is to be in a mental situation that (i) is not a perception of any kind but (ii) cannot be introspectively discriminated from a veridical perception.[13] Some have argued that the indiscriminability view requires a hallucinating subject to have sophisticated conceptual capacities and thus makes it impossible for simple creatures to hallucinate.[14] Whatever the merits of this criticism, it does not apply to the rich pluralist. She says that hallucinations are sensory representations that are mistakenly taken to be perceptions. This can readily occur in simple creatures, since *taking* does not require the capacity to deploy concepts: for a subject to *take* a sensory representation as a perception is just for the sensory representation to play, in that subject, the core functional roles that characterize perception.

The account has one last virtue. Episodic memories and sensory imaginings are not perceptions, but it seems possible, in principle, for the subject to mistakenly *take* them to be perceptions. Thus, the account predicts that it is possible, in principle, for an episodic memory or a sensory imagining to be a hallucination.

I think that this prediction is correct. Consider a war veteran who suffers from flashbacks: he periodically relives the experience of seeing a friend shot in the shoulder. When this happens, he feels as though he is *really there*, watching helplessly as blood pumps from the wound. It is plausible that his episodic memory is, at the same time, a hallucination. Or suppose that when our veteran hears a loud noise, he cannot help but vividly imagine that the people around him are being shot. He gets lost in his imaginings: it is as if he is really seeing these terrible events firsthand. It is plausible that this experience is both a sensory imagining and a hallucination.

Some objections The rich pluralist holds that to be a *hallucination* is to be a sensory representation that is mistakenly taken to be a perception. But couldn't a hallucination fail to be taken to be any way at all? The rich pluralist can reasonably say no: if a sensory representation is not taken to be any way at all—if, for example, the subject is not even *disposed* to treat the content of the representation as true—then this event is just a sensory imagining.

Still, couldn't a subject know that they are hallucinating? The rich pluralist says yes. And wouldn't the subject of a known hallucination fail to take their hallucination to be a perception? The rich pluralist can reasonably say no. Recall that for a subject to *take* a mental item to be of a certain mental kind is for that mental item to play, in that subject, the core functional roles that characterize that mental kind. The rich pluralist holds that a known hallucination still plays the core functional roles of a perception, such as disposing the subject to believe its content. (Again, without this kind of disposition, the subject would just be imagining.) It is just that this disposition is *masked* by the subject's knowledge that they are hallucinating. The objection confuses a masked disposition with no disposition at all.

Conclusion When coupled with their account of perception, the rich pluralist's account of hallucination has some initial explanatory power: it explains what it is for a hallucination to match a perception; it explains why any perception and matching hallucination would cause any subject to form similar beliefs and take similar actions, given a particular mental background; it allows for the possibility of hard-unconscious hallucinations and for the possibility of hallucinations in conceptually unsophisticated creatures; and it correctly predicts that some unusual episodic memories and sensory imaginings are, at the same time, hallucinations.

Still, until I have made a comparison to rival theories, I cannot conclude that these accounts of perception and hallucination are true. My defense of these accounts will be indirect and holistic: over the remainder of the book, I will argue that the total rich pluralist package—including these accounts of perception and hallucination—is superior to all rival theoretical packages.

5.5 Illusion

For reasons of space, I do not discuss illusions in the rest of this book. Still, in the interest of completeness, this section lays out the rich pluralist's theory of illusions, reviews its initial advantages, and responds to an important objection. I will be brief: developing these ideas in detail is a project for another occasion.

I believe that an **illusion** occurs when a subject successfully sensorily represents some particular, *o*; deploys a successful singular representation of some property-instance *F* of *o*; but also misrepresents *F* as being an instance of some sensory quality Q, when in fact it is not. This illusion is *conscious* if, in concert with this, the subject is deeply aware of Q. Suppose, for instance, that a subject consciously sees a scarlet pepper under special lights, so that it looks cobalt blue to them. Then, according to this theory, the subject successfully sensorily represents the pepper and deploys a successful singular representation that locks onto its particular instances of scarlet but misrepresents these as instances of cobalt blue; in concert with this, this subject is deeply aware of cobalt blue (not scarlet).

I believe that this theory has many initial explanatory advantages. I will sketch these quickly, mostly by focusing on the case just described.

First advantage Our subject is directly aware of the particular pepper that they experience. The rich pluralist can explain why: it is because the subject deploys a successful singular representation of that particular pepper.

Second and third advantages On the one hand, our subject is directly aware of the particular instances of scarlet before them; on the other hand, it mistakenly seems to them that these are instances of cobalt blue. The rich pluralist can explain both of these facts. The subject is directly aware of the particular instances of scarlet before them because they deploy successful singular representations *of them*. But it mistakenly seems to them that these are instances of cobalt blue because they misrepresent those instances—the ones that they lock onto via their successful singular representations—as instances of cobalt blue. Structurally, this is like thinking that the Eiffel Tower is in Spain: the concept "the Eiffel Tower" successfully locks onto that particular thing, but the subject goes on to misrepresent how that thing is.

Fourth advantage Like perceptions and hallucinations, illusions can be hard-unconscious. The rich pluralist can explain why: it is because hard consciousness is constituted by deep awareness, and illusions are constituted by sensory representations that can occur in the absence of deep awareness.

Fifth advantage Consider a subject who sees three scarlet peppers. Each is under special lights: the first pepper looks cobalt blue, the second emerald green, and the third canary yellow. Despite being illusory three times over, this experience will reveal to the subject certain essential truths—for instance, that it lies in the essences of these colors that cobalt blue is more similar to emerald green than it is to canary yellow. This is puzzling. How can essential truths about cobalt blue, emerald green, or canary yellow be revealed to a subject who is not in perceptual contact with any instances of those colors?

Rich pluralism answers this question: these essential truths are revealed because the subject is deeply aware of the sensory qualities—the universals—themselves. The details of this explanation were laid out in chapters 2–3.

I will round things out by considering an objection. Return to the subject who sees a scarlet pepper that looks cobalt blue. According to rich pluralism, this subject is deeply aware of cobalt blue. But why is the subject deeply aware of that color *rather than* emerald green or canary yellow? Presumably, it is because the subject *misrepresents* the pepper as being cobalt blue rather than these other colors. Thus ("OI" stands for "objection from illusion"):

OI1. If rich pluralism is true, then the subject is deeply aware of cobalt blue because they misrepresent the scarlet pepper as cobalt blue.

But it makes no sense for the subject to be deeply aware of cobalt blue *because* they misrepresent cobalt blue: a misrepresentation is a case of error, but deep awareness is, by definition, a kind of awareness that does not permit error. Thus:

OI2. The subject is *not* deeply aware of cobalt blue because they misrepresent the scarlet pepper as cobalt blue.

These premises entail:

OI3. Rich pluralism is not true.

My response is that the argument relies on an equivocation. OI1 is true only if the "because" is one of *causal* explanation: the subject's mistakenly representing the pepper as cobalt blue might *cause*, but clearly does not *constitute*, their deep awareness of cobalt blue. However, OI2 is true only if the "because" is one of *constitutive* explanation. It is problematic to say that a misrepresentation, which is in error, might *constitute* deep awareness, which is immune from error. But it is not at all problematic to say that a misrepresentation, which is in error, might *cause* deep awareness, which is immune from error. Thus, the conclusion, OI3, is not supported by the premises.

Conclusion The rich pluralist's theory of illusions has some initial explanatory power and avoids a seemingly powerful objection. I believe that it is the best theory of illusions—but I will not defend that conclusion in this book.

5.6 Intermission

Recall that I am understanding *representationalism* as the view that (i) to perceive is to deploy at least one kind of sensory awareness, typically of external objects and their properties, and (ii) every kind of sensory awareness constitutive of perception is representational (i.e., subject to error). The purpose of the rest of this chapter is to compare rich pluralism to various representationalist alternatives, although some of my observations will apply to other theories, as well.

I have already argued that extended partial revelation is true: in the first instance, all sensory experiences reveal a substantial portion of the essences of the sensory qualities but not of anything else. There are two promising representationalist theories, a monistic one and a pluralistic one, that accept this thesis.

The monistic option is *impure revelatory representationalism*. Like any version of impure revelatory monism, this theory holds that what it is to perceive is to deploy only a single kind of sensory awareness, *sensing*, that targets sensory qualities and particulars but is, in the first instance, only partly revelatory about the essences of the former. The theory adds that sensing is representational—it permits error.

The other option is *pluralistic representationalism*. Like rich pluralism, this theory says that what it is to perceive is to deploy two very different kinds of sensory awareness: partly revelatory awareness of the sensory qualities and nonrevelatory awareness of particulars. However, the pluralistic representationalist thinks that *both* of these kinds of awareness are representational—both permit error. This is in contrast to the rich pluralist, who thinks that only the awareness of particulars is representational; they treat the awareness of sensory qualities as nonrepresentational and thus immune to error.

In the next several sections, I compare these theories to rich pluralism. I first argue that these representationalist options do not have any significant advantages over rich pluralism: all three contenders give equally satisfying accounts of hallucinations (section 5.7); they can all explain the unity of experience (section 5.8); surprisingly, they are all tied in parsimony (section 5.9); and, finally, they can all explain why experience sometimes errs about which sensory qualities are instantiated (section 5.11). But I show that rich pluralism has one crucial advantage: it correctly predicts that experience cannot err about the very essences of the sensory qualities. Its representationalist competitors make the opposite prediction (section 5.10).

5.7 Rich Pluralism versus Representationalism: Hallucinations

The single most important motivation for representationalism is that it can give a satisfying account of hallucinations. In this brief section, I argue that this motivation serves rich pluralism equally well.

Consider a hallucination of a scarlet pepper. This experience will introspectively seem to be just like a perception of a scarlet pepper—in particular, it will seem to make the subject sensorily aware of a particular scarlet pepper. It is hard to understand how this is possible. After all, there are no mind-*independent* peppers around for the subject to be sensorily aware of, and by the framing assumption from chapter 1, neither is the subject sensorily aware of a mind-*dependent* scarlet object.

Representationalists can neatly explain what is happening. This hallucination, they will say, is a *failed* singular sensory representation of a scarlet pepper. In general, failed representations introspectively seem just like their successful counterparts: thus, the hallucination introspectively seems just like a perception of a scarlet pepper. In particular, because the hallucination is a *singular* representation, it seems to make the subject sensorily aware of a *particular* scarlet pepper. Yet a failed singular representation of an *F* can occur even if the subject is not currently related to anything that is *F*. Thus, the hallucination can occur in the absence of any scarlet objects, whether mind independent or mind dependent.[15]

I submit that the rich pluralist has access to the very same account. For they accept all of the usual representationalist posits—in particular, the posit of singular sensory representations, which can be deployed successfully in veridical perceptions or unsuccessfully in hallucinations. Thus, the rich pluralist can help themselves to these same explanations. Of course, the rich pluralist adds *further* posits about deep awareness—nonrepresentational, partly essence-revealing awareness—of sensory qualities. But this does not cause any explanatory interference.

I conclude that the core motivation for representationalism is equally a motivation for rich pluralism.

Next, let us consider a datum that might seem to favor monistic theories over pluralistic ones: the unity of experience.

5.8 Rich Pluralism versus Representationalism: The Unity of Experience

In some obvious sense, conscious experiences seem to be *unified wholes.* They do not seem to be disjointed things but integrated ones.

It might seem that this fact is best explained by monistic theories, such as impure revelatory representationalism. For pluralistic theories, including pluralistic representationalism and rich pluralism, claim that conscious perceptions are constituted by two very kinds of awareness. Would this not predict some kind of disjointedness?

It would not. Consider sentences, which are often constituted by words belonging to many different parts of speech; living organisms, which are often constituted by cells of extremely different kinds; and water molecules, which are always constituted by atoms of very different kinds. Should we conclude that sentences, living organisms, and water molecules are all disjointed things? Of course not: these may still be unified wholes *because of how their diverse parts are organized.* Likewise, conscious perceptions can be unified wholes because they are constituted by diverse kinds of awareness that are organized in the right way.

I conclude that monistic and pluralistic theories alike can explain the unity of experience.

5.9 Rich Pluralism versus Representationalism: Parsimony

In terms of parsimony, *pluralistic* representationalism and rich pluralism are obviously on par. Theorists of both kinds agree that hard-conscious perceptions consist of two kinds of awareness: partly revelatory awareness of sensory qualities and nonrevelatory awareness of particulars. They merely disagree about whether the first kind of awareness is representational or not. However, *impure revelatory* representationalists—indeed, impure revelatory *monists* more generally—might seem to have an edge in parsimony over rich pluralists. For they say that hard-conscious perceptions consist of just one kind of sensory awareness, *sensing,* which targets sensory qualities and particulars alike.

In this section, I will argue that this assessment is incorrect: to account for hard-*un*conscious perceptions, impure revelatory monists must make their theories as complex as rich pluralism.[16]

I begin with this premise ("NAP" stands for "no advantage in parsimony"):

NAP1. If impure revelatory monism is true, then sensing is, by its nature, hard-conscious.

In defense of this premise, suppose for the sake of *reductio* that the impure revelatory monist says that sensing is sometimes hard-conscious (e.g., in ordinary human perception) and sometimes not (e.g., in the hard-unconscious sensory events of honeybees and blindseers). There would then have to be some further factor to make *some* instances of sensing hard-conscious: perhaps hard-consciousness occurs when an instance of sensing is poised to make a direct impact on the belief/desire system or when it is the target of a certain kind of higher-order representation.[17]

This view generates false predictions about when the hard problems of consciousness occur. For, like the rich pluralist, the impure revelatory monist says that we find it hard to understand how purely physical beings could stand in a partly revelatory relation to anything. And they have just added that it is possible to stand in the partly revelatory relation of sensing to the sensory qualities in hard-*unconscious* sensory events. Thus, they predict that we will find it hard to understand how purely physical beings could undergo hard-unconscious sensory events. This prediction is false: we have no trouble understanding how it is possible for a purely physical honeybee to have a hard-unconscious perception of a flower.

This delivers the desired premise: if impure revelatory monism is true, then sensing is, by its nature, hard-conscious.

My next premise is:

NAP2. If sensing is, by its nature, hard-conscious, then hard-unconscious perceptions are constituted by some kind of sensory awareness other than sensing.

To evaluate this premise, suppose that sensing is, by its nature, hard-conscious. Then it can never occur in hard-*un*conscious perceptions—that would make them hard-conscious. So the impure revelatory monist must posit something new to account for hard-unconscious perceptions.

More than this, their new posit must be some kind of *awareness*. For hard-unconscious perceptions, at least the veridical ones, meet all of the criteria for awareness identified in chapter 1: (i) they relate the subject to something—to propositions, for the impure revelatory representationalist; to objects, property-instances, states of affairs, and/or facts, for the impure revelatory naïve realist; (ii) hard-unconscious perceptions are free from error—you cannot veridically perceive a false proposition, a nonexistent object or property-instance, or a nonobtaining state of affairs; (iii) hard-unconsciously perceiving something requires a nonaccidental connection to that thing; and (iv) the category of hard-unconscious perception is a natural kind—it is metaphysically real rather than metaphysically arbitrary.

But isn't it contradictory to posit a form of awareness that is hard-unconscious? No: knowledge is, by definition, a form of awareness, but we can obviously have hard-unconscious knowledge.

I conclude that NAP2 is true: if sensing is, by its nature, hard-conscious, then there must be some new kind of sensory awareness—call it *covert awareness*—that constitutes (veridical) hard-unconscious perceptions.

NAP1 and NAP2 entail:

NAP3. If impure revelatory monism is true, then hard-unconscious perceptions are constituted by some kind of sensory awareness other than sensing.

And now it is clear that impure revelatory monists enjoy no advantage in parsimony over rich pluralists. *Both* theorists now posit two different kinds of awareness. It is just that the rich pluralist thinks that both kinds, deep awareness and successful sensory representation, occur in hard-conscious perceptions, whereas the impure revelatory monist thinks that one kind, sensing, occurs only in hard-conscious perceptions, while the other kind, covert awareness, occurs only in hard-unconscious perceptions.

Conclusion When it comes to parsimony, it is obvious that pluralistic representationalists have no advantage over rich pluralists. What is less obvious is that impure revelatory monists—whether they are representationalists or naïve realists—also enjoy no advantage in parsimony. For, like the rich pluralist, the impure revelatory monist posits two different kinds of awareness; they just distribute them somewhat differently across the cases. This shuffles the cards without changing their number.

So far, then, we are at a deadlock. Impure revelatory representationalists, pluralistic representationalists, and rich pluralists all accept extended partial revelation; thus they can all explain the data from chapters 2–4. In addition, they can all give satisfying accounts of hallucination, they can all explain the unity of experience, and they are equally parsimonious.

But I believe that there is a powerful consideration that breaks this tie.

5.10 Rich pluralism versus Representationalism: Infallibility about Essences

In this section, I argue that, in one very specific respect, experience is infallible. Impure revelatory representationalists and pluralistic representationalists predict otherwise; rich pluralists get things right.

Impure revelatory representationalists, pluralistic representationalists, and rich pluralists all agree that experience can err about particulars—it can misrepresent them. This is how all of these theorists account for hallucinations. However, impure revelatory representationalists and pluralistic representationalists should say that experience can err in a further respect. For these theorists accept extended partial revelation: they think that we have partly essence-revealing awareness of the sensory qualities. And they are representationalists: they agree that *every* kind of awareness constitutive of perception is representational (i.e., subject to error). This entails that partly essence-revealing awareness of sensory qualities is subject to error, as well. Thus:

AI1. If either impure revelatory representationalism or pluralistic representationalism is true, then experience can err about the essences of the sensory qualities.

I do not see how such error is possible. ("AI" stands for "argument from infallibility.") When I reflect on my experiences of scarlet, or pepper-shapedness, or vanilla-scentedness, it seems impossible to me that my experiences could err about the very *essences* of these sensory qualities. For my experience does not just make the relevant essential claims seem *brutely true*. It provides me with an *intelligible basis*, something that lets me understand *why* these claims are true. On reflection, I just cannot see how this intelligible basis could lead me astray, as long as I am being rational. Thus:

AI2. Experience cannot err about the essences of the sensory qualities.

These premises entail:

AI3. Neither impure revelatory representationalism nor pluralistic representationalism is true.

How might my opponents respond to this argument?

Their most natural move is to deny AI2: they might hope to attribute a number of errors to experience. Recall that in chapter 2, I considered a narrower thesis, infallibility, which is effectively the version of AI2 that is restricted to conscious *perception*. I argued that infallibility was intuitively appealing, and I rebutted several objections to it. But let us consider the matter more carefully.

First, my opponents might observe that illusions are perfectly common: a straight stick might look bent in water. This can plausibly be understood as a misrepresentation of the sensory qualities of the stick.

I agree with this assessment but deny that the case meets our requirements. AI2 says that experience never errs about the *essences* of the sensory qualities. Yet it is no part of the essence of bentness whether particular sticks are straight or bent.

Second, my opponent might suggest that some experiences *do* misrepresent the very essences of the sensory qualities: for instance, an experience of supersaturated red, which represents an impossible shade of red, or the waterfall illusion, which represents something as both moving and still.

I do not find these examples convincing. Perhaps it is nomologically impossible for anything to be supersaturated red. But, just on the basis of introspection, I find it very plausible that supersaturated red is a real property, one that it is instantiated in some metaphysically possible worlds. Experience seems to lay bare the character of supersaturated red—it provides an intelligible basis for understanding how that quality is. I find it obvious that something *could* instantiate that quality: my experience gives me a substantive conception, and a perfectly consistent one, of what that would require.

Nor is the waterfall illusion the right kind of case. Grant for the sake of argument that the subject of this illusion represents a single particular as both moving and still, which is metaphysically impossible. Still, this does not suggest that the experience errs about the *essences* of motion or stillness. Quite the opposite: undergoing the waterfall illusion seems to reveal, with perfect accuracy, part of *what it is* to move and to be still. The experience even seems to reveal its own oddity: by itself, the experience provides an intelligible basis for understanding *why* a single entity cannot simultaneously move and be still.

There is one last gambit that the representationalist might try against AI2. They might start by suggesting that the essences of some sensory qualities have been discovered by science. To work with a concrete suggestion, and letting "EE" stand for "experiential error":

EE1. Scarlet—understood as the sensory quality that we encounter in experience—is a surface reflectance property.

However, they might add, experience tells us that scarlet is structurally, and perhaps substantively, nothing like a surface reflectance property. Thus:

EE2. If experience cannot err about the very essences of the sensory qualities, then scarlet—understood as the sensory quality that we encounter in experience—is not a surface reflectance property.

EE1 and EE2 entail:

EE3. Experience can err about the very essences of the sensory qualities.

I regard this as the best argument that my opponents might offer against AI2.

Still, in the end, I think that EE1 and EE2 cannot both be true. Experience seems to reveal that it lies in the essence of scarlet that if something is scarlet, then it is a certain way—call that way *S*. So consider:

Possible instantiation: It is metaphysically possible for something to be *S*.

I accept this claim: I believe that it receives extremely strong support from experience itself. After all, experiences of scarlet give us a clear and substantial conception of *S*. Thus, as long as there is nothing *inherently contradictory* in this conception of *S*—and obviously there is not—possible instantiation will hold. It will hold even if science tells us that reality contains nothing that is *actually S*, for it will remain obvious that there *could* be such things. (By analogy, even if science tells us that reality contains no *actual* perfect Euclidean circles, it is obvious that there *could* be perfect Euclidean circles.)

Once we accept possible instantiation, we are left with two possibilities. One possibility is that being essentially *S* is perfectly compatible with being a surface reflectance property. In this case, EE2 is false: even if experience cannot err about the very essences of the sensory qualities—even if being scarlet requires being essentially *S*—scarlet could still be a surface reflectance property.

Another possibility is that being essentially *S* is *not* compatible with being a surface reflectance property. In that case, there are two distinct properties: a sensory quality, whose essence *requires* being *S*, and a surface reflectance property, whose essence *excludes* being *S*. In that case, EE1 must be false: scarlet—understood as the sensory quality that we encounter in experience—is *not* a surface reflectance property. Science might tell us all about surface reflectance properties, but it does not thereby tell us anything about scarlet. But how could this be? Well, perhaps the sensory quality of scarlet is instantiated in the actual world, but it is distinct from any surface reflectance property—even if the instantiation of scarlet is somehow nomologically or metaphysically necessitated by the instantiation of a certain surface reflectance property. Or perhaps scarlet is just never instantiated in the actual world: perhaps experience makes us aware of a universal that exists but whose only instances are merely possible ones.[18]

Regardless of which of these possibilities obtains, experience cannot err about the essences of the sensory qualities, just as AI2 says. I conclude that this premise is true.

Now return to AI1, which says that if impure revelatory representationalism or pluralistic representationalism is true, then experience sometimes errs about the essences of the sensory qualities. The impure revelatory representationalist, in particular, might hope to reject this premise. After all, we saw in chapter 3 that *any* impure revelatory monist (whether representationalist or naïve realist) should accept the following claims:

S1. By its nature, sensing reveals, or at least purports to reveal, at least part of how its targets are intrinsically.

S2. It is always a priori that if a sensory quality (or instance thereof) is intrinsically a certain way, then it is *essentially* intrinsically this way.

S3. It is never a priori that if an object, an event, or a high-level property is intrinsically a certain way, then it is *essentially* intrinsically this way or even essentially this way at all.

The impure revelatory representationalist uses S1–S3 to explain *one* asymmetry: that experience is *partly essence-revealing* for the sensory qualities and instances thereof, but not for objects, events, or high-level properties.

So they might hope to use these same claims to explain another, seemingly similar asymmetry: that experience *cannot err* about the essences of the sensory qualities and instances thereof, even though it is perfectly fallible about objects, events, and high-level properties.

I have trouble seeing how this stratagem could succeed. The problem is that the impure revelatory representationalist's theory seems to support precisely the opposite conclusion from what they want. After all, the impure revelatory representationalist thinks that sensing is representational. Thus, given S1, they should think that, among other things, sensing *represents* the intrinsic characters of the sensory qualities and their instances. However, there is no particular obstacle to *misrepresentation* here: it is perfectly possible to represent scarlet as being intrinsically *G*, where *G* is in fact the intrinsic character of lime green. So suppose that sensing sometimes has a false content such as this. According to S2, it is then a priori that if scarlet is intrinsically *G*, then scarlet is *essentially* intrinsically *G*. Thus, it should be possible for the subject to conclude on the basis of their experience that scarlet is essentially intrinsically *G*. In this sense—the only sense, given this theory, in which this experience comments at all on the essence of scarlet—the experience would be in error.

I conclude that AI1 is true: if impure revelatory representationalism or pluralistic representationalism is true, then experience sometimes errs about the essences of the sensory qualities. From AI1 and AI2, I deduce AI3: neither impure revelatory representationalism nor pluralistic representationalism is true.

Still, let us consider one last argument for representationalism over rich pluralism.

5.11 Rich Pluralism versus Representationalism: Fallibility about Instantiation

I have been arguing that our awareness of the sensory qualities is immune to error and is thus nonrepresentational. The representationalist might resist this conclusion. They might concede that experiences never err about *the essences of the sensory qualities*. Still, they might argue, they can at least err about *whether the sensory qualities are instantiated*. Thus, our awareness of the sensory qualities must be representational after all.[19]

Let us inspect this argument more closely. The representationalist begins with this premise:

FI1. Experience can err about whether the sensory qualities are instantiated.

("FI" stands for "fallibility about instantiation.") I grant this premise: like the representationalist, I think that a hallucination of a scarlet pepper is in error about whether scarlet is instantiated.

Next, my opponent observes that, by definition, deep awareness is immune to error. Thus, they assert:

FI2. If experience can err about whether the sensory qualities are instantiated, then, in experience, we represent the sensory qualities rather than being deeply aware of them.

These premises entail:

FI3. In experience, we represent the sensory qualities rather than being deeply aware of them.

I would resist FI2. For suppose that experience can err about whether the sensory qualities are instantiated. It follows that *the part of experience that permits this error* must be representational. But it does not follow that *every part of experience* must be representational. In other words, what is true is not FI2, but:

FI2*. If experience can err about whether the sensory qualities are instantiated, then *the part of experience that permits this error* is representational.

FI2* is entirely compatible with rich pluralism. For the rich pluralist holds that what permits such errors are the subject's *sensory representations* of various sensory qualities as being instantiated. Deep awareness does not, by itself, permit error on this topic (or any other), for it is, by itself, silent about whether the sensory qualities are instantiated.

I conclude that not only representationalists, but also rich pluralists, can explain why experience is fallible about whether the sensory qualities are instantiated.

It is instructive to combine the lesson of this section with that of the previous section. I have observed that we need to explain (i) the respect in which experience is *fallible* (namely, about *whether the sensory qualities are instantiated*) and (ii) the respect in which experience is *infallible* (namely, about the *essences* of the sensory qualities). I have argued that while representationalists and rich pluralists can both explain (i), only rich pluralists can explain (ii). That is an important fragment of my case for rich pluralism over representationalism.

5.12 Rich Pluralism versus Representationalism: The Final Verdict

I have been understanding *representationalism* as the view that (i) to perceive is to deploy at least one kind of sensory awareness, and (ii) every kind of sensory awareness constitutive of perception is representational (i.e., subject to error). We now have the materials for seeing why rich pluralism is superior to any form of representationalism. All we must do is assemble them.

We can begin by dividing representationalist theories into three categories. First there is *nonrevelatory representationalism,* which denies extended partial revelation. My arguments against this theory are simply my arguments for extended partial revelation, given in chapters 2–4. The remaining two representationalist theories accept extended partial revelation: *impure revelatory representationalism* is the best monistic version of this theory, while *pluralistic representationalism* is the best pluralistic version. Both of these theories predict that perception can err about the very essences of the sensory qualities. However, this is impossible. Rich pluralism delivers the correct prediction.

I conclude that rich pluralism has important advantages over any representationalist alternative. However, before we can give our final verdict, we must consider whether there are any important advantages in the other direction.

We can start by considering the core explanatory benefits of representationalism. The representationalist can explain why perceptions make us sensorily aware of particulars: it is because perception is at least partly constituted by successful singular sensory representations. Relatedly, the representationalist can explain why hallucinations seem to make us sensorily aware of particulars while actually failing to do so: it is because hallucinations are the failed counterparts of perceptions; they are failed singular sensory representations. Finally, the representationalist can explain why experience can err about whether the sensory qualities are instantiated: it is because experience *represents* sensory qualities as instantiated and can thus *misrepresent* them as instantiated. The rich pluralist accepts these representationalist posits and thus enjoys these same explanatory benefits.

Next, we want to explain why conscious experiences are unified wholes. Presumably, monists can explain this. But so can pluralists: plenty of unified wholes, such as sentences, living organisms, and water molecules, consist of diverse parts that are organized in the right way.

We must also consider parsimony. Obviously, the pluralistic representationalist has no advantage here over the rich pluralist. Less obviously, neither does the impure revelatory representationalist. They are monists, accounting for conscious perception via just one kind of awareness, *sensing*. However, they grant that sensing is partly essence-revealing; thus, they cannot yet account for hard-*un*conscious awareness, which is not at all essence-revealing. Their best resort is to posit a new, *covert* form of awareness. This makes their theory just as complex as rich pluralism.

The upshot is that the explanatorily adequate representationalist theories do not have any important advantages over rich pluralism, but there is one major advantage running in the opposite direction.

I conclude that rich pluralism is better than any representationalist alternative. That is a central thesis of this part of the book, part II.

5.13 Looking Forward

It remains to be seen how rich pluralism compares to its naïve realist rivals. To be sure, I have made *some* remarks about this. My arguments for extended partial revelation rule out versions of naïve realism that reject this thesis. In addition, I observed in this chapter that adequate versions of impure revelatory monism, *including naïve realist versions*, will have no advantage in parsimony over rich pluralism. However, I have not yet engaged with any of the core motivations for naïve realism. That is the central task of the next part of the book: part III, comprising chapters 6–9.

Once that is done, I can make my final comparison between naïve realism and rich pluralism.

III Rich Pluralism versus Naïve Realism

I have been understanding *naïve realism* as the view that (i) to perceive is to deploy at least one kind of sensory awareness, typically of external particulars and their properties, and (ii) every kind of sensory awareness constitutive of perception is nonrepresentational (i.e., immune to error) in the sense identified in chapter 5. We can restate the second part of the view by reserving the label *acquaintance* for any nonrepresentational relation of sensory awareness. Then, what (ii) says is that every kind of sensory awareness constitutive of perception is a form of acquaintance.

In this part of the book, comprising chapters 6–9, I will make a detailed comparison between rich pluralism and various versions of naïve realism.

Although naïve realism has been propped up by countless sophisticated arguments, my sense is that few of these are what really drive most naïve realists. For the most part, were these arguments revealed to be unsound, naïve realists would simply look for new ones. However, there are four ideas that I think have a different status. The first idea is that hard-conscious perception reveals the intrinsic characters of its targets. The second is that hard-conscious perception positions us to make a special kind of knowledgeable *de re* reference to its targets. Third, there is the idea that hard-conscious perceptions *present* their targets, in a certain sense. The fourth and final idea is that hard-conscious perceptions *introspectively seem* to present their targets. For each of these ideas, many have thought that naïve realism provides the only adequate account. These are, to my mind, the truest motivations for naïve realism.

In this part of the book, I will knock over these motivations. My contention will be that all four of these ideas can be fully honored by rich pluralists (and, for that matter, by representationalists). Chapter 6 focuses on revelation of intrinsic characters, chapter 7 shifts to knowledgeable *de re* reference and presentation, and chapter 8 takes up apparent presentation. Recall also that adequate versions of naïve realism make no gains in parsimony, as I concluded from my discussion of impure revelatory monism in chapter 5.

This leaves us with no strong reasons to prefer naïve realism to rich pluralism. But, as I will argue in chapter 9, there are good reasons for the reverse preference: to account for the full range of sensory events—including hallucinations, hard-unconscious perceptions, and sensory imaginings—naïve realists must greatly complicate their theories.

On the basis of this holistic comparison, I will conclude that rich pluralism is superior to any naïve realist alternative.

6 Awareness of Intrinsic Characters

When I see a round tomato, I become aware of the *intrinsic character* of what I perceive: this is a rough statement of the *intrinsicality datum* (section 6.1). The central question of this chapter is how best to express, and explain, this datum.

It is often believed that the intrinsicality datum can only be explained by naïve realists: this is the first major motivation for naïve realism. I regard this as a misconception. It has arisen because the datum has not been stated in its most general form.

First, there is a problem of *scope*: previous theorists have not correctly identified *which* experiences reveal the intrinsic characters of things. Once we get the scope of the datum right, it becomes clear that many naïve realist theories are explanatorily inadequate, although a few naïve realist theories are up to the task (sections 6.2–6.3). Then there is a problem of *content*: previous theorists have not correctly identified exactly *what* experience reveals about the intrinsic characters of things. Once this content is properly stated, we see that the intrinsicality datum is beyond the explanatory reach of *all* nonrevelatory theories, including naïve realist ones (section 6.4). What is needed to explain the datum is extended partial revelation. It is irrelevant whether we combine this thesis with naïve realism, representationalism, or rich pluralism (section 6.5).

Against this, some naïve realists have argued that no representationalist theory can do justice to the datum, and these arguments can equally be aimed at rich pluralism. My reply is that these arguments do not undermine representationalist commitments; at best, they undermine *reductive physicalism*. Thus, representationalists and rich pluralists can sidestep the arguments by taking certain facts about experience as primitive (section 6.6).

In the end, then, the intrinsicality datum does not support naïve realism over representationalism or rich pluralism. It supports theories that accept extended partial revelation against theories that reject that thesis, and perhaps it also supports primitivist theories over reductive physicalist ones (section 6.7).

6.1 The Intrinsicality Datum: A First Pass

There is an obvious but hard-to-characterize distinction between *intrinsic* and *extrinsic* properties. *Intrinsic properties* are properties that concern how an entity and its parts are in and of themselves, regardless of any relations to other entities. For example, roundness is an intrinsic property: what makes a particular tomato round is purely a matter of how its parts are arranged internally. *Extrinsic properties* are properties that are not intrinsic (e.g., the property of being within 1 meter of a chair). Extrinsic properties should not be confused with *relational properties*, for some relational properties of an entity are intrinsic to it. Take, for example, the property of having more hair on one's head than one's feet. This is an intrinsic property, since it concerns only how the individual in question is. But it is also a relational property: it concerns a relation of quantity between head hair and foot hair.

Hard-conscious perception plays an important role with respect to intrinsic properties. When you see a round tomato, your experience does more than just make you aware of roundness, a property that is *in fact* intrinsic. It also makes all of the following claims seem true to you:

1. That roundness is an intrinsic property.
2. That anything round must be intrinsically *that way*.
3. That the tomato is intrinsically *that way*.

All of these claims are distinct. 1 says only that roundness is an intrinsic property, without speaking to the specific intrinsic character of round things. Thus, 1 does not entail 2, which does speak to the specific intrinsic character of round things. Nor does 2 entail 1. For having a property might require being intrinsically that way *and* meeting some extrinsic conditions, in which case the property is extrinsic overall: think of the property of being a round object that is within 1 meter of a chair. Finally, 3 is logically independent from 1 and 2. On the one hand, 1 and 2 are silent about the tomato, and thus fail to entail 3, which says something about the tomato. On the other hand, 3 is silent about roundness and thus fails to entail 1 and 2, which are about roundness.

Let us try to express the relationship between perception and claims 1–3 in a more precise way. Recall that for a mental item *M* to *reveal* a truth *p* about some entity *e* is for the following conditions to be met:

(i) Mental event *M* makes the subject aware of entity *e*.

(ii) For a conceptually sophisticated subject, rational reflection on mental event *M* will, by itself, make it seem that *p* is true.

(iii) The subject's awareness of entity *e*, as described in (i), provides an intelligible basis for its seeming that *p* is true, as described in (ii).

What seeing a round tomato does is make you aware of the tomato, and of roundness, in a way that *reveals* to you the truth of claims 1–3. Thus, we can initially state the datum like this:

> The *intrinsicality datum (first pass)*: Hard-conscious perceptions reveal to us that certain perceived properties are intrinsic, that anything with those properties is intrinsically a certain way, and that certain objects are intrinsically a certain way.

Notice that hard-conscious perceptions need not reveal the *entire* intrinsic characters of perceived properties and objects. To repurpose an argument from chapter 2: colors divide into the compound (orange, purple), which are mixtures of multiple other colors, and the elemental (red, green), which are not. Seeing green does not by itself reveal its status as elemental—but it is presumably intrinsic to green that it is an elemental color. Likewise for perceived *objects*: seeing a pepper need not reveal its complete shape. Thus, hard-conscious perceptions need only reveal *a substantial portion* of the intrinsic characters of perceived properties and objects.[1] This is how we should understand the intrinsicality datum.

I will briefly mention that there is a distinct but closely related datum here. To appreciate what it is, distinguish two kinds of properties. There are *dispositional properties*: salt is disposed to dissolve when placed in warm water, and glass is disposed to break when struck. But the instantiation of a dispositional property is ordinarily grounded in the instantiation of one or more properties that are not dispositional but rather *categorical*. The instantiations of the dispositions just mentioned, for instance, are grounded in the instantiations of certain categorical microstructural properties.[2] The distinction is an obvious one, although, as with many obvious distinctions, it is difficult to characterize it precisely.

Categoricity has sometimes been confused with intrinsicality.[3] This should be avoided. For one thing, what makes a property categorical is, roughly, being nondispositional, but what makes a property intrinsic is, roughly, not involving relations to wholly distinct entities. In addition, categoricity and intrinsicality are extensionally distinct. Consider the property, which I currently instantiate, of being less than a meter from my laptop screen. This property is categorical rather than dispositional, but it is also evidently extrinsic rather than intrinsic to me. Moreover, there could be intrinsic dispositions: dispositions that something has purely in virtue of how it and its parts are—for example, dispositions for an entity's parts to interact with each other in particular ways.

Just as hard-conscious perception reveals certain truths about intrinsicality, it also reveals certain truths about categoricity. For instance, seeing a round tomato reveals that roundness is not only an intrinsic property but also a categorical one, that anything round is categorically *that way*, and that the tomato is categorically *that way*. Thus, we also have:

> The *categoricity datum (first pass)*: Hard-conscious perceptions reveal to us that certain perceived properties are categorical, that anything with those properties is categorically a certain way, and that certain objects are categorically a certain way.[4]

The intrinsicality datum and the categoricity datum are so similar that they will presumably be amenable to parallel explanations. Thus, in the interest of space, I will focus purely on explaining the intrinsicality datum.

It has influentially been argued that naïve realism is the only theory that can explain the intrinsicality datum. Is this true?

6.2 The Intrinsicality Datum Generalized

In this section, I will show that naïve realists can indeed explain the first-pass statement of the intrinsicality datum. But I will argue that this statement is incomplete: the intrinsicality datum applies not only to hard-conscious perceptions but also to *all* hard-conscious experiences, including hard-conscious hallucinations. I will argue in the next section that this undermines many naïve realist theories of hallucination.

Recall that *naïve realism* is the view that (i) to perceive is to deploy at least one kind of sensory awareness, typically of external objects and their properties (and/or property-instances), and (ii) every kind of sensory awareness constitutive of perception is nonrepresentational—it is a form of *acquaintance*. Although all current naïve realists appear to be monists, I have tailored this definition to allow for the possibility of pluralistic naïve realism.

Our current explanatory target is:

> The *intrinsicality datum (first pass)*: Hard-conscious perceptions reveal to us that certain perceived properties are intrinsic, that anything with those properties is intrinsically a certain way, and that certain objects are intrinsically a certain way.

Many naïve realists propose to explain this by adding just one more posit, namely:

> **Acquaintance reveals intrinsic character (ARIC):** Every acquaintance relation constitutive of hard-conscious perception reveals a substantial portion of the intrinsic character of its targets.[5]

Here, *revealing a substantial portion of the intrinsic character of x* is to be understood as revealing that *x* is intrinsic and/or that *x* is intrinsically a certain way.

It is an interesting question for the naïve realist why ARIC would be true. Perhaps it is because when a subject is acquainted with an object or property, they are not only sensorily aware of it; it is, in some interesting sense, a *constituent* of the subject's experience.[6] Or perhaps ARIC is simply a primitive fact about acquaintance. Either way, the naïve realist will say that seeing a round tomato is a matter of being acquainted with the tomato and the property of roundness (or an instance of it). According to ARIC, this reveals a substantial portion of the intrinsic character of that property and of the tomato. So far, so good.

But there is a problem lying in wait.

Consider an unusual subject, Beatrice, who has seen all sorts of objects but, for whatever reason, has never seen anything round. One day, Beatrice merely *hallucinates* something round—a round tomato, perhaps. (She has encountered plenty of tomatoes before, but they have been roughly cubical. Perhaps Beatrice lives in a world in which tomatoes have undergone significant genetic mutations.) Beatrice's hallucination will not reveal anything about "the" tomato—there is no such thing. However, her hallucination *will* reveal something about roundness—in particular, that roundness is an intrinsic property and that anything round is intrinsically a particular way. In these respects, hallucination is on all fours with perception. Had Beatrice been *perceiving* a round tomato rather than *hallucinating* one, this would not in any way have improved her appreciation of the intrinsic character of roundness.

The same could be said of other kinds of hard-conscious experiences: a vivid sensation of a round afterimage, or a vivid sensory imagining of a round object. Just like hard-conscious perceptions, these reveal a substantial portion of the intrinsic character of roundness. Thus, a better way of stating the datum is this:

> The *intrinsicality datum (second pass)*: Hard-conscious experiences of any kind—including hard-conscious perceptions, hallucinations, sensations, sensory imaginings, and episodic memories—reveal that certain *properties* are intrinsic and that anything with those properties is intrinsically a certain way. Hard-conscious perceptions also reveal that perceived *objects* are intrinsically a certain way.

This poses an explanatory challenge. For naïve realists usually think that acquaintance occurs only in hard-conscious perception, never in hard-conscious hallucination (or in sensations and sensory imaginings). Thus, as of yet, such theories contain nothing to explain why hard-conscious hallucinations, sensations, and so on reveal any portion of the intrinsic characters of things.

Many naïve realists will be tempted to simply reject the datum: they will be tempted to say that hallucinating a round tomato does not *really* reveal a substantial portion of the intrinsic character of roundness but merely *seems* to do so. Call this the *error theory*. I believe that it is a poor choice. We can appreciate why by making a holistic comparison between error-theoretic versions of naïve realism and rich pluralism.

The intrinsicality datum, as it applies to hallucination, is extremely initially plausible. This point should not be glossed over: take a moment to reflect carefully on a hallucination of a round tomato. It just seems perfectly obvious that this will genuinely *reveal* to you a substantial portion of the intrinsic character of roundness. Thus, the error theorist loses a lot of credit for denying this claim.

But they might hope to offer compensation for this. In particular, the error theorist might suggest that only naïve realism can fully honor the four motivating ideas about hard-conscious perception discussed in the introduction to this part of the book: first, the idea that hard-conscious perception reveals a substantial portion of the intrinsic characters of things; second, the idea that hard-conscious perception positions us to make a special kind of knowledgeable *de re* reference to its targets; third, the idea that hard-conscious perception presents its targets, in a certain sense; and fourth, the idea that hard-conscious perception (correctly) introspectively seems to present its targets.

However, the arguments from this chapter to chapter 9 are intended to show that the rich pluralist can honor all of those ideas about perception, while *also* honoring many important intuitions about hard-conscious hallucinations—for instance, the intuition that hard-conscious hallucinations reveal a substantial portion of the intrinsic characters of various properties. (Later chapters will turn up additional intuitions about hard-conscious hallucinations that error theorists must deny.) If these arguments are correct, then rich pluralism has all of the benefits of naïve realism but without the costs of the error theory. In the final analysis, this makes rich pluralism the clear winner.

Thus, I think that it is wiser for the naïve realist to try to genuinely *explain* the second-pass version of the intrinsicality datum. How might they do so?

6.3 Some Naïve Realist Explanations

In this section, I will run through many familiar naïve realist theories of hallucination. I will argue that most of them cannot explain why the intrinsicality datum applies to hallucinations, although a few can.

The indiscriminability view According to this view, what it is to hallucinate is to be in a mental situation that is not a perception of any kind but cannot be first-personally discriminated from a veridical perception.[7] Here, for two situations to be *indiscriminable* for a subject is for them to be unable to know which of these situations they are in. Thus, the indiscriminability theorist thinks that a defining characteristic of hallucination is the subject's in-principle *ignorance* that they are not veridically perceiving. I do not see how occupying a mental situation constituted by in-principle ignorance could reveal a substantial portion of the intrinsic character of roundness.[8]

Some naïve realists have replied that lacking this discriminatory ability can position the subject to gain at least one piece of positive knowledge: knowledge that *either* they are perceiving (say) a round object *or* it merely seems to them that this is so.[9] The indiscriminability theorist might suggest that being in a position to have this positive knowledge is what reveals a substantial portion of the intrinsic character of roundness.

Let us examine this view further. Consider a mathematician, Sandra, who knows the geometrical definition of roundness and is fluent in using this knowledge to deduce various truths about round things. But Sandra is blind, so her understanding is purely theoretical: she has no idea what a round thing (or a point, or a line) really is.[10] Suppose that Sandra is told by a reliable source that *either* her friend James is currently perceiving something round *or* it merely seems to him that he is currently perceiving something round. Call this claim *p*. On the basis of this testimony, Sandra is in a position to know *p*. Evidently, however, this does not reveal to her any portion of the intrinsic character of roundness. Thus:

IV1. Sandra's being in a position to know that <either James is currently perceiving something round or it merely seems to him that he is currently perceiving something round> does *not* reveal to Sandra any portion of the intrinsic character of roundness.

("IV" stands for "indiscriminability view.") However, the indiscriminability theorist asserts that:

IV2. Sandra's being in a position to know that <either she herself is currently perceiving something round or it merely seems to her that she is currently perceiving something round> *does* reveal to Sandra a substantial portion of the intrinsic character of roundness.

It is curious to endorse these claims together. The situations described in IV1 and IV2 differ only in whether the subject knows something about someone else or herself. How would that make a difference to what is revealed to her about the intrinsic character of roundness, when this property has nothing to do with subjects?

In sum, the indiscriminability theorist might say that hallucination reveals a substantial portion of the intrinsic character of roundness simply because to hallucinate is just to be in-principle ignorant—to lack a discriminatory ability. That claim is immediately implausible. Or she might say that hallucination reveals a substantial portion of the intrinsic character of roundness because it positions the subject to know that <either she is currently perceiving something round, or it merely seems to her that she is currently perceiving something round>. But then the indiscriminability theorist is committed to a curious asymmetry, since knowing a similar proposition about someone else does not, by itself, reveal any portion of the intrinsic character of roundness.

I conclude that the indiscriminability theorist lacks a good explanation of the intrinsicality datum.

The cognitive view According to this view of hallucination, what it is to hallucinate a round tomato, for example, is to undergo a mental event that lacks "phenomenal character"[11] but that produces the same cognitive effects that a veridical perception of that object would have produced in a rational subject with the same background mental situation.[12] Can any of these cognitive effects explain the intrinsicality datum?

Veridically seeing a round tomato would certainly produce many *beliefs*, *desires*, and the like. For example, it might produce the belief that the subject is seeing the shiny red surface and leafy green top of a tomato and the desire to eat the tomato. However, in the case of a veridical perception, it is not these subsequent beliefs and desires, but the experience itself, that reveals a substantial portion of the intrinsic character of roundness. Thus, the same should be true in the hallucinatory case: by themselves, these beliefs and desires should not reveal any portion of the intrinsic character of roundness.

Of course, there is a further effect that veridically seeing a round tomato would have on a rational subject: it would reveal to them a substantial portion of the intrinsic character of roundness. But the cognitive theorist cannot appeal to *this* effect—the very question at hand is how this effect is possible in the hallucinatory case. The cognitive theorist might claim that having this revealed is part of what *makes it the case* that the subject is hallucinating. But if the cognitive theorist cannot explain how this could be revealed to the subject to begin with, then their account just ends up entailing that it is impossible to hallucinate.

Here is another way to put the problem. We can all agree that for a subject in a certain background mental situation, a neural event that does not even partly constitute a perception might *just happen* to have in them many of the effects that a perception would have had. The cognitive theorist says that any such neural event is a hallucination. Fine! But if one effect of a neural event is revelation of the intrinsic character of roundness, then the cognitive theorist still owes us an account of how a neural state could have *that* effect.

Thus, the cognitive theorist cannot explain the intrinsicality datum.

The mental object view According to this view, what it is to hallucinate is to be acquainted with a mental object.[13] I grant that if the naïve realist accepts the mental object view, then they can explain the datum: they can say that a subject who hallucinates a round tomato is acquainted with a mental object that is genuinely round, and acquaintance reveals a substantial portion of the intrinsic characters of its targets (ARIC). However, this view violates the framing assumption from chapter 1: it says that in hallucination, we are sensorily aware of peculiar mental entities. Thus, I set it aside.

Interlude So far, we have been examining naïve realist theories of hallucination that either cannot explain the intrinsicality datum at all or can only explain it by accepting very costly commitments. I shift now to some naïve realist theories of hallucination that are more explanatorily adequate.

Flexible naïve realism The flexible naïve realist holds that sensory qualities like redness are *ontologically flexible*: the existence of an instance of redness can be fully grounded *either* in the existence of an ordinary object like a tomato *or* in a subject's being in a certain internal state (perhaps a brain state). Thus, the flexible naïve realist says that when I hallucinate a red tomato, I am aware of a perfectly real but *mind-dependent* instance of redness. The existence of this instance of redness is fully grounded in the fact that I am in a certain internal state, and the instance does not inhere in anything—not even my mind or experience. But what happens when I really see a red tomato? Then I am aware of an instance of redness whose existence is *ontologically overdetermined*: its existence is fully grounded in the existence of my internal state *and, separately*, in the existence of the tomato.[14]

Technically speaking, flexible naïve realism violates the framing assumption from chapter 1. Part of the framing assumption is that perceptions and hallucinations do not make us sensorily aware of peculiar mental entities. However, the flexible naïve realist thinks that, in hallucinations, we are sensorily aware of instances of sensory qualities that—peculiarly—are not only purely mind dependent but also fail to inhere in anything. Thus, we could justly disqualify flexible naïve realism from the race. But I will not do this: I find it instructive to compare flexible naïve realism to other forms of naïve realism and to rich pluralism.

In fact, a major advantage of flexible naïve realism is that it permits a satisfying explanation of the intrinsicality datum. The flexible naïve realist can say that even in hallucination, the subject is genuinely acquainted with properties like redness and roundness. They can add that acquaintance reveals a substantial portion of the intrinsic character of its targets (ARIC).

The imaginative view According to this view, part of what it is to be a hallucination is to be an involuntary sensory imagining. This is not intended as a full account of the nature of hallucination: it is granted that some involuntary sensory imaginings are not hallucinations.[15]

Perhaps the partial account is enough to explain the datum: my opponent need only add that sensorily imagining a round tomato reveals to the subject the intrinsic character of roundness. But this just pushes back the problem: it remains to be explained how sensory imagination can do this.

A promising response for the imaginative theorist is that sensory imaginings are representations *of perceptions*.[16] As a naïve realist, the imaginative theorist might think that they already have an account of how perceptions of round tomatoes reveal to us the intrinsic character of roundness. They might add that in virtue of being representations of perceptions, sensory imaginings have the same revelatory power. To be sure, not all representations of perceptions are revelatory in this way: a mere thought about a perception does not reveal the intrinsic character of anything. But sensory imaginings should be understood as *distinctively sensory* representations—representations that somehow borrow the revelatory power of the perceptions they represent.

This account leaves much to be filled in. Still, I grant that in principle, the imaginative theorist can explain the intrinsicality datum.

The quality-awareness view According to this view, what it is to perceive is to be acquainted with particulars instantiating sensory qualities, understood as universals, and what it is to hallucinate is just to be acquainted with sensory qualities, *without* being acquainted with any particulars.[17] This gives the naïve realist a satisfying explanation of the intrinsicality datum: they can say that hallucinating a round tomato is a matter of being acquainted with sensory qualities like roundness and that acquaintance reveals a substantial portion of the intrinsic characters of its targets (ARIC).

I am satisfied with this explanation of the intrinsicality datum.

Pluralistic naïve realism The pluralistic naïve realist posits two very different kinds of acquaintance relations. The first, call it *shallow acquaintance*, provides sensory awareness of particulars and is not revelatory in any way. The other, call it *deep acquaintance*, provides sensory awareness of the sensory qualities, understood as universals, *and* reveals a substantial portion of their intrinsic characters (and perhaps also their essences). The idea is that shallow acquaintance, acquaintance with particulars, occurs only in hard-conscious perceptions, but deep acquaintance, acquaintance with sensory qualities, occurs in hard-conscious perceptions *and* hard-conscious hallucinations.

The pluralistic naïve realist can say that in cases of hallucination, the subject is deeply acquainted with sensory qualities; this reveals their intrinsic characters. This is an effective explanation of the intrinsicality datum.

Conclusion The naïve realist can easily explain the first-pass version of the intrinsicality datum, which concerns only hard-conscious perceptions. However, properly understood, the datum extends to *all* hard-conscious experiences, including hard-conscious hallucinations. Can the naïve realist account for this?

One option is to accept an error theory, which says that hallucinations do not *actually* reveal any portion of the intrinsic characters of the sensory qualities; they merely *seem* to do so. I suggested that, in the final analysis, this approach will be inferior to rich pluralism: the rich pluralist not only honors all of the core motivations for naïve realism but also explains the intrinsicality datum, as well as many other important intuitions about hallucination. However, this evaluation will not be fully substantiated until the end of chapter 9.

A wiser option for the naïve realist is to grant the intrinsicality datum and try to explain it. I argued that the indiscriminability view and the cognitive view cannot explain the datum at all; meanwhile, the mental object view can explain the datum but is unacceptable on other grounds. The best *monistic* naïve realist options are flexible naïve realism, the imaginative view, and the quality-awareness view; *pluralistic* naïve realism is also satisfactory.

I conclude that, of all the naïve realist theories developed so far, these four are the best options.

6.4 The Intrinsicality Datum Refined; a Problem for Any Nonrevelatory Theory of Perception

In this section, I refine the intrinsicality datum further to arrive at my final statement of it. I argue that this version of the intrinsicality datum rules out any *nonrevelatory* theory of perception—any theory that rejects the existence of any kind of essence-revealing awareness—whether that theory is naïve realist, representationalist, or something else.

Earlier, I have argued that seeing a round tomato, or for that matter hallucinating one, reveals that roundness is an intrinsic property and that anything round is intrinsically a particular way. I now observe that these experiences reveal something more: that all of this *lies in the very essence of roundness*. Reflect on an experience of roundness and you will see what I mean: this, by itself, will teach you that roundness is *essentially* an intrinsic property and that it lies in the *essence* of roundness that anything round is that way. What is more, an experience of roundness does not just make these claims seem brutely or inexplicably true. It provides an *intelligible basis* for extracting these truths.

To this, I add a contrasting observation. A tomato plausibly has *some* of its intrinsic properties essentially, such as the intrinsic property of being made of matter. However, although my experience of a tomato will inform me of some of its intrinsic properties, it will be silent about whether any of those properties are essential to the tomato: we saw as much in chapter 3.

By assembling these observations, we arrive at:

> The *intrinsicality datum (final version)*: Sensory experiences of any kind—including hard-conscious perceptions, hallucinations, sensations, sensory imaginings, and episodic memories—reveal that it lies in the essences of certain *properties* that they are intrinsic and that anything with those properties is intrinsically a certain way. In addition, hard-conscious perceptions reveal that perceived *objects* are intrinsically a certain way but not whether they are essentially that way.

I believe that this is the most precise and comprehensive statement of the datum.

It is at this point that *any* nonrevelatory theory of perception, whether representationalist, naïve realist, or something else, runs into a problem. By definition, a nonrevelatory theory rejects the existence of any kind of essence-revealing awareness. This is flatly incompatible with the intrinsicality datum, which, in its final form, identifies particular essential truths that are revealed in experience.

A nonrevelatory theorist might respond by objecting to this final statement of the intrinsicality datum. They might say that these essential truths—that it lies in the essence of roundness that roundness is an intrinsic property and that anything round is intrinsically a certain way—are not actually *revealed* by my experience. Surely what my experience immediately serves up is something less complicated than that.

This objection misunderstands the definition of revelation. That a truth is revealed does not entail that it is "immediately served up": an experience counts as revealing a truth as long as the truth seems to be true purely on the basis of rational reflection, *no matter how complex or extended,* on the experience. This is just what we find: you can learn that it lies in the essence of roundness that roundness is an intrinsic property and that anything round is intrinsically a certain way, purely by reflecting on an experience of roundness. The objection to nonrevelatory theories stands.

Conclusion Naïve realists have often worked with an incomplete understanding of the intrinsicality datum and thus have missed its true significance. One omission is that it is not only hard-conscious perceptions but *all* hard-conscious experiences that reveal a substantial portion of the intrinsic characters of the sensory qualities. Many naïve realists have trouble explaining why this is the case, although flexible naïve realists, imaginative theorists, quality-awareness theorists, and pluralistic naïve realists are in a better position. A second omission is that hard-conscious experiences do not just reveal that roundness is intrinsic and is intrinsically a certain way. They reveal that all of this lies in the very *essence* of roundness. This is flatly incompatible with any nonrevelatory theory of perception, be it naïve realist, representationalist, or something else.

Still, even if *not all* naïve realists can explain the true intrinsicality datum, you might think that *only* naïve realists can explain it. For only they posit acquaintance, which, by ARIC, is what reveals the intrinsic characters of things. Is this reasoning correct?

I think not.

6.5 The Intrinsicality Datum Explained via Extended Partial Revelation

According to many naïve realists, what is ultimately required to explain the intrinsicality datum is *acquaintance* (i.e., *nonrepresentational awareness*). I disagree: I think that what we must ultimately posit is *partly essence-revealing awareness*. Whether this awareness is representational or not is irrelevant. Thus, I think that a satisfying explanation of the intrinsicality datum is available to any theorist who accepts extended partial revelation.

In this section and the next, I defend these claims.

What we want to explain is:

> The **intrinsicality datum**: Sensory experiences of any kind—including hard-conscious perceptions, hallucinations, sensations, sensory imaginings, and episodic memories—reveal that it lies in the essences of certain *properties* that they are intrinsic and that anything with those properties is intrinsically a certain way. In addition, hard-conscious perceptions reveal that perceived *objects* are intrinsically a certain way but not whether they are essentially that way.

I suggest that the following thesis can carry us most of the way:

> *Extended partial revelation*: In the first instance, all hard-conscious experiences reveal a substantial portion of the essential truths about the sensory qualities and nothing else.

The intrinsicality datum is about hard-conscious experiences in general, not just hard-conscious perceptions. Extended partial revelation explains this by saying that partial revelation of essences occurs in *all* hard-conscious experiences. In addition, according to the datum, hard-conscious experiences reveal not only that certain truths about intrinsicality obtain but also that they lie within the *essence* of roundness. Again, extended partial revelation explains this: it says that, in general, experience reveals a substantial portion of the very *essences* of the sensory qualities. Finally, the datum says that experience does *not* reveal whether the intrinsic properties of the tomato are essential to it. Once more, extended partial revelation explains why: this occurs because experience does *not* reveal any essential truths about objects.

This gives us another argument for extended partial revelation: the thesis provides an extremely natural explanation of many aspects of the intrinsicality datum.

However, by itself, extended partial revelation does not explain the intrinsicality datum in its entirety. In particular, it does not explain why hard-conscious experiences reveal that *objects* are intrinsically a certain way. We have considered a number of theories that accept extended partial revelation: rich pluralism, impure revelatory representationalism/naïve realism, and pluralistic representationalism/naïve realism. Fortunately, each of these theories has a way to pick up the explanatory slack.

Start with the rich pluralist. They offer the following posits about deep awareness:

DA1. Deep awareness occurs in all *sensory experiences*, including all hard-conscious perceptions, hallucinations, sensations, sensory imaginings, genuine episodic memories, and merely apparent episodic memories.

DA2. At least in the sensory experiences of beings like us, the only targets of deep awareness are certain universals—the *sensory qualities*.

DA3. Deep awareness is a nonrepresentational form of awareness-of that reveals to the subject a substantial portion—not necessarily all!—of the essences of its targets.

The rich pluralist uses these posits to explain the truth of extended partial revelation and thus gains the explanatory power of that thesis. However, they also posit a second form of awareness, *successful sensory representation*, that targets particulars and occurs in hard-conscious perceptions and episodic memories. This lets the rich pluralist explain the remainder of the intrinsicality datum. They can say that when I see a round tomato, I deploy a successful sensory representation that makes me aware of it, while also being deeply aware of roundness. I then put these two forms of awareness together: I represent the tomato *as* having this property, roundness. That is why my experience reveals that the *tomato* is intrinsically round.

Pluralistic representationalists and naïve realists can give effectively the same explanation, insofar as they agree with the rich pluralist that perception is constituted by two very different kinds of awareness, one partly revelatory, the other not. To be sure, these theorists disagree with the rich pluralist about exactly which forms of awareness are representational and which are not. However, this is immaterial: the rich pluralist's explanation does not turn on the representational or nonrepresentational character of either kind of awareness.

Now consider impure revelatory monism. According to this theory, not only hard-conscious perceptions, but all hard-conscious experiences, involve a single relation of sensory awareness. This relation, call it **sensing**, can target both particulars and sensory qualities. However, there is something special about the sensory qualities: when they are sensed, their essences are partly revealed. By contrast, sensing a particular does not reveal any part of its essence. More formally, the impure revelatory monist says this:

IRM1. What it is to have a hard-conscious experience is to enjoy exactly one kind of sensory awareness: *sensing*.

IRM2. In hard-conscious perceptions and episodic memories, the subject senses particulars as having sensory qualities. In all other hard-conscious experiences, the subject senses sensory qualities alone.

IRM3. Sensing is a relation of sensory awareness that reveals a substantial portion of the essences of the sensory qualities but never reveals anything about the essences of particulars.

IRM3 lets the impure revelatory monist accept extended partial revelation, which in turn lets them explain much of the intrinsicality datum, as described above. What they still need to explain is why hard-conscious perceptions reveal that *objects like tomatoes* are intrinsically a certain way. Their answer is simple: seeing a round tomato partly consists of sensing roundness, which reveals certain ways that that property is, intrinsically. But seeing a round tomato also partly consists of sensing *the tomato itself* and sensing it *as* round. This reveals that the *tomato* is intrinsically round.

It is worth recalling that impure revelatory monism is equally available to representationalists and naïve realists. A crucial difference between these two versions of impure revelatory monism is in how they handle hallucinations. The representationalist will say that in a hallucination of a round tomato, the subject deploys an unsuccessful and nonrevelatory *sensory representation* that fails to lock onto any particular tomato but a successful and partly essence-revealing *sensory representation* of roundness. The naïve realist might say that in the hallucinatory case, the subject is *nonrepresentationally* aware of roundness in a partly revelatory way. The naïve realist does not want to appeal to tomato representations, so they will need to provide some other kind of explanation for why the subject seems to be aware of a tomato. Still, as far as explaining the intrinsicality datum goes, both of these versions of impure revelatory monism seem perfectly satisfactory.

Conclusion What is required to explain the intrinsicality datum is not the idea that experiential awareness is *nonrepresentational* but the idea that it is *partly essence-revealing (for the sensory qualities)*. More precisely, what is required is extended partial revelation. Thus, the intrinsicality datum can be effectively explained by all of theories that accept this thesis, including rich pluralism, pluralistic representationalism/naïve realism, and impure revelatory representationalism/naïve realism.

There is another moral to draw. In the previous section, we found that if a theory posits no essence-revealing awareness whatsoever, then it cannot explain the intrinsicality datum in its full generality. In this section, we made the complementary discovery that the intrinsicality datum *can* be explained in full by all of the major theories that accept extended partial revelation. This gives us yet another argument for extended partial revelation.

Looking forward We are not out of the woods yet. It has been argued that certain representationalist commitments—which are shared by representationalists and rich pluralists—make it impossible to explain the intrinsicality datum. My task in the next section is to evaluate these challenges.

6.6 Some Challenges to the Explanatory Adequacy of Representationalism and Rich Pluralism

I have argued that rich pluralists, pluralistic representationalists, and impure revelatory representationalists can all explain the intrinsicality datum in full. (So can pluralistic naïve realists and impure revelatory naïve realists, but that will not be relevant here.) In all three cases, part of the explanation is that hard-conscious perceptions involve sensory representations of objects *as* having intrinsic properties. However, some have argued that this cannot be right: sensory representations cannot possibly latch onto intrinsic properties. They can only latch onto dispositional and extrinsic ones.

In this section, I will examine several arguments built around this core concern. I will show that the proper target of these concerns is not a commitment to sensory representations but a commitment to reductive physicalism. Thus, all it takes to avoid the concern is to adopt a certain form of *primitivism*.

One version of the argument runs as follows.[18] Assume, for the sake of *reductio*, a purely representationalist account of object perception:

A1. To hard-consciously perceive objects is simply to deploy sensory representations that represent those objects via a perceptual relation. (Assumed for *reductio*.)

My opponent then claims that:

A2. The content of any sensory representation is determined purely by (i) what causes the representation and (ii) what functional role the representation actually plays, or ideally would play, in the subject's mind.

This is just intended to express the commitments of a typical representationalist theory of content.

The next premises are:

A3. Any sensory representation will be causally affected in exactly the same way by any objects with exactly the same dispositional and extrinsic properties. Any further differences in an object's underlying intrinsic properties are irrelevant.

A4. What functional role a sensory representation actually plays, or ideally would play, in the subject's mind is determined purely by the dispositional and extrinsic properties that an object has. Any further differences in the object's underlying intrinsic properties are irrelevant.

We can understand the thought here by comparing a normal subject seeing a round tomato (call her Nicky) with a subject who has recently had her brain placed in a vat (call her Vicky). We may suppose that the brains of Nicky and Vicky are intrinsically identical at any given time. But whereas the brain state underlying Nicky's perceptual experience is caused by an ordinary tomato, the brain state underlying Vicky's perceptual experience is caused by electrical impulses sent by wire from a computer. As long as the tomato and the computer wire have the same extrinsic dispositions, they will have exactly the same effects on their respective subjects, and it will be rational for the subjects to respond in exactly the same ways. This, despite the fact that the tomato and the wire have radically different intrinsic characters.[19]

Let us connect these threads. A1 tells us that to perceive objects is simply to deploy sensory representations that represent those objects via a perceptual relation, and A2 says that the content of these representations is determined purely by a certain causal relation and functional role. A3 and A4 inform us that neither the causal relation nor the functional role is affected by an object's intrinsic properties; only its dispositional and extrinsic properties make any difference. These premises strongly suggest (albeit without entailing):

A5. To perceive objects is to represent them as having only dispositional and extrinsic properties.

However, a central observation of this chapter has been that perception reveals to us a substantial portion of the intrinsic characters of objects. Thus:

A6. To perceive objects is not to represent them as having only dispositional and extrinsic properties.

A5 and A6 contradict each other. Thus, we must reject our starting assumption, A1. We must conclude that:

A7. To perceive objects is not simply to deploy sensory representations that represent those objects via a perceptual relation.

There ends the argument.

I think that this argument contains two important mistakes. The first mistake is in this stretch of the argument:

A2. The content of any sensory representation is determined purely by (i) what causes the representation and (ii) what functional role the representation actually plays, or ideally would play, in the subject's mind.

A3. Any sensory representation will be causally affected in exactly the same way by any objects with exactly the same dispositional and extrinsic properties. Any further differences in an object's underlying intrinsic properties are irrelevant.

A2 entails that *one* factor in determining the content of a sensory representation is what causes the representation. This, together with A3, is supposed to show that *this* particular factor does not generate categorical or intrinsic content. However, this is a bad inference. The problem is that A3 does not directly comment on what causes sensory representations; it only identifies conditions under which distinct causes would affect a sensory representation in the same way. This is an entirely different matter.

We can appreciate the difference by examining the premise that the anti-representationalist really needs, which is:

A3*. Intrinsic properties are never what cause sensory representations.

A3* is clearly false. We can see this by returning to Nicky, our normal subject who sees a tomato. I grant that, in principle, something with radically different intrinsic properties, such as a computer emitting electrical impulses, might have had just the same narrow causal effects on any putative sensory representations in Nicky's brain. But it is perfectly clear that, given the actual circumstances, the tomato's intrinsic properties, such as its shape, *do* cause Nicky's sensory representations: had the tomato been different in shape, Nicky's sensory representations would have been different in corresponding ways.[20]

Still, this leaves us in aporia. Nicky's sensory representations are caused by *both* the tomato's categorical, intrinsic properties, such as its shape, *and* its merely dispositional, extrinsic properties. It would seem redundant to include both sets of properties in the content of Nicky's sensory representations. Which set of properties should we prefer?

This brings me to my second criticism of the argument. I think that the representationalist should also reject:

A2. The content of any object representation is determined purely by (i) what causes the representation and (ii) what functional role the representation actually plays, or ideally would play, in the subject's mind.

I encourage the representationalist to accept the popular view that content is also partly determined by a third factor: *naturalness*. The idea is that an interpretation is better to the extent that it assigns referents that are more natural—more metaphysically real, more structural, more joint-carving.[21] Now, the categorical, intrinsic properties of an entity are arguably much more natural than its dispositional, extrinsic properties. Thus, the representationalist can reasonably say that the categorical, intrinsic properties, and *not* the dispositional, extrinsic properties, are the ones that make it into the content.

For these reasons, I think that this anti-representationalist argument is unsound.

However, I think that there is a simpler and more compelling argument along the same lines. The argument begins with this premise:

B1. Hard-conscious experiences reveal a substantial portion of the intrinsic characters of objects.

As we saw in the opening sections of this chapter, B1 is clearly true. Part of the idea was that hard-conscious experiences do not just seem to make us aware of *truths* about the intrinsic characters of sensory qualities. They seem to provide us with the *intelligible basis* for extracting those truths.

The next premise is:

B2. If representationalism is true, then the content of any sensory representation is determined purely by (i) what causes the representation; (ii) what functional role the representation actually plays, or ideally would play, in the subject's mind; and (iii) the naturalness of any candidate contents.

This is intended as an uncontroversial description of representationalist theories of content.

Finally, we have:

B3. If the content of any sensory representation is determined purely by (i) what causes the representation; (ii) what functional role the representation actually plays, or ideally would play, in the subject's mind; and (iii) the naturalness of any candidate contents, then hard-conscious experiences do not reveal a substantial portion of the intrinsic characters of objects.

This premise has a lot of intuitive plausibility. Suppose that sensory representations have their contents determined purely by factors (i)–(iii). Then, as we have seen, sensory representations might have contents about properties that are *in fact* intrinsic. These contents might even make it rational for us to believe that these properties are intrinsic and are intrinsically a certain way—even essentially so. But it is hard to see how such truths could be *revealed* to us—how experience could make us aware of these properties in a way that provides an *intelligible basis* for extracting these truths. In other words, the concern is that, given the representationalist's theory of content, it is very difficult to see how they could accept extended partial revelation—but this thesis is crucial for explaining the intrinsicality datum.

Together, B1–B3 entail:

B4. Representationalism is not true.

What should we make of this argument?

One option is to reject B3—to hold, in effect, that partly revelatory awareness can be given a purely physical/functional reduction. I confess that I am genuinely uncertain whether B3 is true. For I am genuinely uncertain whether hard consciousness can be given a purely physical/functional reduction, and as I argued in chapter 4, partly revelatory awareness is what constitutes hard consciousness. Thus, I will grant B3 for the sake of inquiry.

Must we then reject representationalism? Certainly not. We can still reject B2, which says that if representationalism is true, then the content of any sensory representation is determined purely by (i) what causes the representation; (ii) what functional role the representation actually plays, or ideally would play, in the subject's mind; and (iii) the naturalness of any candidate contents. This premise might seem very plausible if we focus exclusively on versions of representationalism that give *reductive analyses* of representation in purely physical/functional terms. But the representationalist is not forced to accept such an account. They can instead introduce *primitive facts* about how sensory representations get their contents. Call this view *primitivist representationalism*.[22]

Primitivist representationalism is surely a legitimate view. After all, it is fair game for the naïve realist to posit a primitive relation of acquaintance that *just does* reveal a substantial portion of the intrinsic characters of particulars and/or sensory qualities. Thus, it must be fair game for the representationalist to posit a primitive kind of representation that *just does* reveal a substantial portion of the essences of the sensory qualities, as per extended partial revelation. And once the primitivist representationalist has taken extended partial revelation on board, they can accept impure revelatory monism, which—as we saw in the previous section—lets them explain the intrinsicality datum in full.

Thus, the current line of argument might undermine reductive versions of representationalism, but it poses no threat to primitivist representationalism.

The argument turned against rich pluralism Even if the primitivist representationalist can avoid the argument, the rich pluralist posits sensory representations *without* taking them as primitive. Thus, they still owe a response to a very similar argument. It begins with the same idea as before, namely:

C1. Hard-conscious experiences reveal a substantial portion of the intrinsic characters of objects.

The next step is:

C2. If rich pluralism is true, then the content of any sensory representation is determined purely by (i) what causes the representation; (ii) what functional role the representation actually plays, or ideally would play, in the subject's mind; and (iii) the naturalness of any candidate contents.

This premise is eminently reasonable: the rich pluralist wants to accept some kind of naturalistic theory of representation, and C2 offers a wide umbrella for such theories.

The last premise is:

C3. If the content of any sensory representation is determined purely by (i) what causes the representation; (ii) what functional role the representation actually plays, or ideally would play, in the subject's mind; and (iii) the naturalness of any candidate contents, then hard-conscious experiences do not reveal a substantial portion of the intrinsic characters of objects.

C1–C3 entail:

C4. Rich pluralism is not true.

There concludes the argument.

Fortunately, the rich pluralist has good reason to reject C3. All that is true, the rich pluralist can say, is that if the content of any sensory representation is determined purely by factors (i)–(iii), then *sensory representations alone* do not reveal a substantial portion of the intrinsic characters of objects. That does not show that *hard-conscious experiences* do not reveal a substantial portion of the intrinsic characters of objects, because hard-conscious experiences do not consist solely of sensory representations. They also consist of instances of deep awareness, which are perfectly capable of revealing the intrinsicality of certain *properties*. The subject can then combine their deep awareness with their object representation to represent the *object* as having these intrinsic properties. Thus, deep awareness and sensory representation *together* reveal a substantial portion of the intrinsic characters of objects.

But wait—how does deep awareness manage to reveal a substantial portion of the intrinsic characters of the sensory qualities? If deep awareness is meant to be analyzed in a reductive physicalist way, then the problem returns. Just as it is hard to see how *representations*, understood in a reductive physicalist way, could genuinely reveal anything, it is hard to see how *anything*—say, deep awareness, *or even a naïve realist's nonrepresentational relation of acquaintance*, could genuinely reveal anything, *if this awareness relation is understood in a reductive physicalist way*.

The moral is that the concern, even if it is sound, does not undercut representationalism while raising up nonrepresentationalism. Instead, it undermines reductive physicalist views of any kind—including reductive physicalist versions of naïve realism—while raising up primitivist views. This makes it clear how the rich pluralist can respond: rather than accepting a reductive physicalist analysis of deep awareness, they can just take deep awareness as primitive.

Conclusion Our hard-conscious experiences do not just make it *seem* that certain properties are essentially intrinsic and are essentially certain intrinsic ways. Our experiences *reveal* these facts to us. The challenge is to explain how this is possible. Contrary to what some naïve realists have thought, this challenge cuts across the representationalism/nonrepresentationalism divide. If the challenge is sound—and I am genuinely uncertain whether it is—then it undermines *all* reductive physicalist theories. However, we can surmount the challenge by treating partly revelatory awareness as primitive, *regardless* of whether that awareness is constituted by representations, deep awareness, or acquaintance.

That move, together with an acceptance of extended partial revelation, gives us a fully satisfying explanation of the intrinsicality datum.

6.7 Conclusion

In the introduction to this part of the book, I identified four core motivations for naïve realism. The first motivation was that hard-conscious perceptions reveal a substantial portion of the very intrinsic characters of objects, and naïve realism is, allegedly, the only theory that can explain how this is possible. The purpose of this chapter has been to sap this motivation.

I began by stating the datum more precisely. For one thing, revelation of intrinsic characters does not occur only in hard-conscious perceptions; it occurs in *all* hard-conscious experiences. In addition, what experience reveals is not only that *objects* have a certain intrinsic character but that various *properties* have certain intrinsic characters and indeed essentially so. Thus, the true datum is this:

> The *intrinsicality datum*: Sensory experiences of any kind—including hard-conscious perceptions, hallucinations, sensations, sensory imaginings, and episodic memories—reveal that it lies in the essences of certain *properties* that they are intrinsic and that anything with those properties is intrinsically a certain way. In addition, hard-conscious *perceptions* reveal that perceived *objects* are intrinsically a certain way but not whether they are essentially that way.

I then considered how different theorists might explain this datum. Part of the datum is that even hallucinations reveal the intrinsic characters of sensory qualities. I argued that many versions of naïve realism struggle to explain this, although flexible naïve realism, the imaginative view, the quality-awareness view, and pluralistic naïve realism are up to the task. Another part of the datum is that experience reveals certain truths about the *essences* of certain sensory qualities. This, I argued, cannot be explained by any nonrevelatory theory of perception, whether naïve realist, representationalist, or something else: by definition, such theories say that perception never reveals any essential truths to us. I went on to show that what it takes to explain the intrinsicality datum in full is, primarily, to accept extended partial revelation—which, incidentally, bolsters the case for that thesis. This explanation is available to naïve realists, representationalists, rich pluralists, and strange pluralists.

I then considered the concern that insofar as representations get their contents in certain ways, they cannot genuinely *reveal* intrinsic characters. I showed that this concern is not properly directed against representationalist theories but against reductive physicalist theories of all kinds—including even reductive versions of naïve realism. Still, all of these theorists can avoid the concern by taking partly revelatory awareness as primitive.

The upshot is this. The intrinsicality datum does not support naïve realism over representationalism, as is usually thought. It actually undermines nonrevelatory versions of both theories. What the intrinsicality datum really favors are theories that accept extended partial revelation and perhaps also theories that take partly essence-revealing awareness as primitive.

In the next chapter, I discuss two more motivations for naïve realism.

Appendix 6A: Campbell on Awareness of Intrinsic and Categorical Characters

A central datum of this chapter, stated very roughly, is that experience reveals to us how objects are intrinsically. This claim is inspired by ideas from John Campbell's book *Reference and Consciousness* (2002), so it is important to show that I have actually engaged with Campbell's thinking, as it is most charitably understood. That is the purpose of this appendix.

Campbell's discussion betrays some serious confusions, which makes the interpretive task a difficult one. One problem is that Campbell speaks over and over of *objects* as being categorical (pp. 138–145). However, only *properties* can be categorical. What does Campbell mean, then? There are several clues.

One clue is that Campbell thinks that the categoricity of objects explains why they tend to transmit marks over time:

> [First, a]n object at a later time is capable of bearing marks transmitted by the way it was earlier. . . . Suppose that while at school you carve your initials on a desk. When you revisit years later, there they still are. In this case, the identity of the object over time seems to be the categorical ground of its potentiality to sustain this kind of marking interaction. The reason why the marks are there is that it's the same desk. . . . We need the notion of a categorical object, whose categorical identity can be the ground of its complex of dispositions to interact with other objects. (p. 140)

A second clue is that Campbell thinks that the categoricity of objects explains why they tend to have a range of correlated effects at a given time:

> The second type of phenomenon in explaining which we use the identity of macroscopic objects has to do with the tendency of macroscopic objects to produce ranges of correlated effects. For example, one person addressing a group of people may produce a range of effects in them: there may be a shared sense of lassitude, or on the other hand intense intellectual excitement, after such an encounter.

> What explains the correlation in the effects produced is, in part, that it was one and the same person addressing all these people. (p. 140)

A third clue is that Campbell consistently elides the distinction between the intrinsic and the categorical. For example, he holds that "experience is experience of the categorical," but he calls this the *Intrinsicness Condition* (2002, 137). Similarly, he says this:

> We would ordinarily regard shape properties as the paradigmatic *categorical* properties of objects. Roundness is the reason why the thing tends to roll if suitably propelled and so on; it is the *intrinsic* ground of the complex functional state described. (p. 139, emphasis added)

I propose the following interpretation: Campbell's most fundamental thought is that perception positions us to think of objects as having certain *intrinsic properties*. This would explain why Campbell focuses on the tendencies of objects to transmit marks over time and to have a range of correlated effects at a given time: the intrinsic properties of an object can frequently explain these tendencies. At the same time, many of the intrinsic properties of objects—perhaps all of their most explanatorily fundamental ones and certainly the most paradigmatic ones—are categorical. This might explain why Campbell runs together intrinsicality and categoricity.

But there is a further interpretive problem: Campbell repeatedly says that perception positions us to conceive of the *identity* of an object—see his (2002, 138–145)—but it is unclear what he means by this. Everything is identical to itself, so what is the difference between being confronted with the identity of an object and being confronted with the object itself? Perhaps Campbell (sometimes) has in mind the identity of an object across time—that is, perhaps he means that perception positions us to conceive of objects as persisting over time.

Yet this reading runs into problems. For Campbell seems to think of the categoricity and identity of an object as one and the same: "How is it that we have this conception of the identity of a macroscopic object, as the categorical ground of these dispositions?" (p. 141). But whether an object persists over time is orthogonal to whether it has categorical properties. Consider a demon who creates a table, immediately destroys it, and then creates an intrinsically identical table in the same location, over and over again. Every table in this series has categorical properties, but none of these tables persists over time.

Perhaps this problem can also be solved if we interpret Campbell as focusing, most fundamentally, on the intrinsic character of objects. For we can then read him as thinking that *haecceities*—which distinguish qualitatively identical entities—are intrinsic. In that case, we can read Campbell as emphasizing that seeing a table over time reveals another intrinsic aspect of that object: the fact that it is the *same* object that persists over time.

I conclude that on the most charitable interpretation of Campbell, the crucial datum about perception is that it reveals that perceived objects are intrinsically a certain way. I have indeed engaged with this idea: it is part of all of my formulations of the intrinsicality datum in this chapter.

7 Strong Referential Character

At the beginning of this part of the book, I identified four core motivations for naïve realism. The previous chapter dealt with the first of these motivations. It is now time to engage with the second and third ones.

The second core motivation for naïve realism is that hard-conscious perceptions provide a special kind of knowledge of reference, and supposedly, only naïve realists can explain this fact. I begin my examination of this motivation by showing that the datum is not limited to hard-conscious perceptions. It extends, at least partially, to *all* experiences (section 7.1). Once the datum has been stated crisply, it is natural for the representationalist to try to explain it simply by attributing successful singular content to experiences. However, I show that this account is inadequate (section 7.2). I then consider the *intrinsic awareness account*—pioneered by naïve realists—which says that experience provides knowledge of reference in virtue of revealing the intrinsic characters of things. I show that with some important refinements, this account gives us a satisfying account of the datum (section 7.3). However, the intrinsic awareness account is not the exclusive property of naïve realists. It can be claimed by representationalists and rich pluralists, too, leaving no one with any comparative advantage (section 7.4).

Then there is the third core motivation for naïve realism. This is that all and only hard-conscious perceptions involve the distinctively valuable relation of **presentation** (section 7.5). Again, some have thought that naïve realists have a uniquely effective account of this—indeed, one that is just an elaboration of the intrinsic awareness account. Naïve realists have not quite found the best version of the account, but I argue that, with some good tailoring, their account can be made to fit the datum snugly (section 7.6). Then I show that this account, too, is perfectly available to representationalists and rich pluralists (section 7.7), objections notwithstanding (section 7.8).

Thus, the second and third motivations for naïve realism are washed away. The naïve realist is right to think that we should explain knowledge of reference, as well as presentation, by saying that experiences reveal the intrinsic characters of things. Where they go wrong is in thinking that naïve realism has a proprietary claim on this idea. In truth, the idea is available to all of the major monistic and pluralistic theories of perception (section 7.9).

7.1 Strong Referential Character Introduced

Naïve realists have observed that experience provides a special kind of knowledge of reference. Let us see what this means.

Consider a man who hears various sounds coming from the neighboring house. On that basis, he posits the existence of various objects that are likely in the house—a coffee table, a refrigerator, and so on. His experiences have what I will call **super-weak referential character** toward these targets: these experiences position him to make reference to these objects. Suppose that later the man learns from his neighbors that the house does indeed contain a coffee table, one that is made of espresso-colored wood. The man's experience of hearing this information arguably has what I will call **weak referential character** toward the coffee table: it positions him to make *de re* reference to the coffee table.

One day, however, the man enters the house and *sees* the coffee table. Now his experience has what I will call **strong referential character** toward it: not only does the experience position him to make *de re* reference to that target, but it also positions him to *know, in a special way, what he is referring to*. We may add that the (complete) strong referential character of a mental item is its maximal property of the form, *positioning the subject to make knowledgeable reference to. . . .* In calling this property *maximal*, I mean that the ellipses are to be filled in with *all* applicable targets.[1]

Super-weak referential character is cheap. A subject can make reference to an entity just on the basis of knowing what functional roles the entity performs: a detective might refer to *that criminal*—the person, identity unknown, who left this bootprint, shattered that window, knocked over this lamp, and so on. Even weak referential character is cheap. The ability to make *de re* reference can be granted by the perceptual relation, even in cases of super-blindsight: a subject can make *de re* reference to a chair that they super-blindly see. What is precious, and much harder to account for, is strong referential character.

Which mental items have strong referential character, and toward what?

Some philosophers have observed that *hard-conscious perceptions* have strong referential character toward *external objects and property-instances*.[2] When I see a scarlet pepper, not only can I make *de re* reference to the pepper and its instances of scarlet; I can also know what I am thereby referring to. It is also plausible that hard-conscious perceptions have strong referential character toward universals. Seeing a scarlet pepper positions me to make knowledgeable *de re* reference to scarlet itself—to appreciate what that property is *in general*.

These thinkers have also observed that hard-*un*conscious perceptions lack strong referential character entirely. Consider a super-blindseer or a zombie whose visual perceptions are entirely hard-unconscious. Perhaps these perceptions still position the subject to make *de re* reference to various targets. However, it is clear that the subject will not know what they are thereby referring to, in the relevant sense.[3]

It is usually thought that this is the end of the story. I disagree: I find strong referential character in other cases, too.

Suppose that you remember seeing a scarlet pepper, conjuring a crisp image of it in your mind. Just like hard-conscious perceptions, your hard-conscious episodic memory will position you to make *de re* reference to the pepper, its instances of scarlet, and the universal scarlet, *and to know what you are thereby referring to*.

Now suppose that you merely hallucinate a scarlet pepper. This time, your experience will not have strong referential character toward "the" pepper—there is no pepper. But it will plausibly have strong referential character toward the universal scarlet: you will be able to know, in the relevant sense, *what scarlet is*. For similar reasons, a vivid sensation of scarlet will not have strong referential character toward any particulars, but it will plausibly have strong referential character toward the universal scarlet.[4] It is possible to develop an error theory that denies these initially plausible claims. But it is a cost, in and of itself, to take that approach. For reasons described in the last chapter, I think that this cost is ultimately not worth paying.

In short, it seems that *all* experiences have strong referential character toward *sensory qualities*, but only hard-conscious *perceptions and episodic memories* have strong referential character toward *particulars*. Meanwhile, all hard-unconscious mental items lack strong referential character entirely.

How can we explain these facts?

7.2 A Simple Representationalist Account

In this section, I will examine the most obvious representationalist account of strong referential character and argue that it is inadequate. Part of the framing assumption from chapter 1 is that perception makes us sensorily aware of *particulars*; thus, I will assume that the representationalist posits singular content. It should be obvious that the problem would cut even deeper for the representationalist who posited only existential contents.

The singular representationalist can acknowledge an important difference between perceptions and matching hallucinations. They can say that perceptions have *successful* singular contents: the content of a perception of a scarlet plate will be partly constituted by a particular plate and some particular instances of scarlet. That will not be true of a matching hallucination. Notice that to say that the *content* of an experience is partly constituted by one or more particulars is not yet to say that the *experience itself* is partly constituted by one or more particulars. The representationalist might or might not wish to accept this further commitment.

This suggests the following account:

> The **content constitution account**: For a mental item to have strong referential character toward some target *T* is for it to have a content that contains *T* as a constituent.[5]

This account has some important virtues. For one thing, it correctly predicts that hard-conscious perceptions will have strong referential character toward any particulars that the subject perceives. In fact, given the plausible assumption that hard-conscious *episodic memories* have successful singular contents, the account correctly predicts that these, too, will have strong referential character toward particulars. Finally, the representationalist can say that all experiences have contents that contain some sensory qualities as constituents and thus have strong referential character toward them.

Still, the content constitution account is clearly inadequate. Many other contentful mental states or events have contents that are partly constituted by particulars—for instance, super-blindsight and singular thoughts. These mental items might have *weak* referential character: they might position the subject to make *de re* reference. But they do not have *strong* referential character: they do not position them to *know what they are thereby referring to*, in the special sense at issue. Thus, the account explains only part of the phenomenon, and the easier part at that.[6]

Having a successful singular content might suffice for weak referential character, but it does not suffice for strong referential character. We need a better account.

7.3 The Intrinsic Awareness Account of Strong Referential Character

In this section, I consider an alternative account given by some naïve realists. I argue that it is inadequate as it stands but can be tailored to fit the datum snugly.

What some naïve realists have proposed is this ("IR" stands for "intrinsicality" and "reference"):

The *intrinsic awareness account of strong referential character (first pass)*:

IR1*. For a mental item to position the subject to make knowledgeable reference to some target *T*—equivalently, for it to have *strong referential character* toward *T*—is for it to position the subject to make *de re* reference to *T and* reveal a substantial portion of the intrinsic character of *T*.

IR2*. All and only hard-conscious perceptions reveal a substantial portion of the intrinsic characters of their targets, including particulars such as objects and property-instances (and perhaps also universals).[7]

This account has its merits. It entails that all hard-conscious perceptions have strong referential character toward some particulars, which is true. The account also entails that all hard-*un*conscious mental items lack strong referential character of any kind, which is also true.

But the account has severe flaws. To begin with, it predicts that hard-conscious episodic memories, sensory imaginings, hallucinations, and sensations will entirely lack strong referential character. As we have seen, this is implausible: *all* experiences plausibly have strong referential character toward sensory qualities, and hard-conscious episodic memories also have strong referential character toward particulars. A closely related problem is that the last chapter contained direct arguments against (IR2*): yes, *all* hard-conscious perceptions reveal a substantial portion of intrinsic characters, but it is not *only* hard-conscious perceptions that do that. Experiences of any kind will at least reveal a substantial portion of the intrinsic characters of the sensory qualities.

In light of these problems, I would refine the account as follows:

The *intrinsic awareness account of strong referential character (final version)*:

IR1. For a mental item to position the subject to make knowledgeable reference to some target *T*—equivalently, for it to have *strong referential character* toward *T*—is for it to position the subject to make *de re* reference to *T and* reveal a substantial portion of the intrinsic character of *T*.

IR2. Hard-unconscious mental items never reveal any portion of the intrinsic characters of anything.

IR3. All hard-conscious mental items—that is, experiences—make the subject aware of sensory qualities in a way that reveals a substantial portion of their intrinsic characters.

IR4. Some experiences—including all hard-conscious perceptions, all hard-conscious episodic memories, and some hard-conscious sensory imaginings—also make the subject aware of particulars *as* instantiating, or being instances of, sensory qualities.

IR5. Some experiences—including all hard-conscious sensations, as well as hard-conscious hallucinations that are purely of novel objects—do not make the subject directly aware of any particulars.

All of these claims enjoy independent plausibility. We will return shortly to IR1, but IR2 and IR3 are directly supported by the arguments of the previous chapter. Meanwhile, IR4 is immediately plausible: it is very plausible that seeing or remembering a scarlet pepper makes the subject aware of the pepper and its instances of scarlet. Similarly for IR5: hallucinating a scarlet pepper or having a sensation of a scarlet afterimage obviously does not make the subject directly aware of any mind-*independent* particulars, so at best, these experiences can make the subject directly aware of mind-*dependent* particulars—sense-data. But the framing assumption from chapter 1 rules out this possibility.

Now return to IR1. This claim speaks of *revealing a substantial portion of the intrinsic character of a target*. This should be understood as playing out differently for targets of different kinds. For a *universal*, revealing a substantial portion of its intrinsic character is understood straightforwardly; for an *object*, revealing a substantial portion of its intrinsic character is a matter of revealing a substantial portion of the characters of its intrinsic properties; for a *property-instance*, it is a matter of revealing a substantial portion of the character of the corresponding universal.

The central motivation for IR1 is that once a subject can make *de re* reference to something, it is revelation of a substantial portion of its intrinsic character that positions them to know, in the relevant sense, what they are referring to. Revelation of intrinsic character "lights up" the referent for them. In this way, the intrinsic awareness account identifies, in a transparent and precise way, what is special about this kind of knowledge of reference.

Thus, the posits of the intrinsic awareness account are individually plausible. What is more, when taken collectively, they explain the data.

One datum is that super-blindsight, zombie perceptions, and abstract singular thoughts do not have any strong referential character whatsoever. We can explain this by observing that these are all hard-unconscious. Hard-unconscious mental items never reveal any portion of the intrinsic characters of anything (IR2), but revelation of intrinsic characters is required for strong referential character (IR1).

A second datum is that all experiences have strong referential character toward some sensory qualities. To explain this, we can say that all experiences make the subject aware of sensory qualities (IR3), thereby positioning the subject to make *de re* reference to those qualities. This occurs in a way that reveals a substantial portion of the intrinsic characters of those sensory qualities (IR3 again). Together, these features suffice for strong referential character (IR1).

A third datum is that if they are hard-conscious, then all perceptions, all episodic memories, and some sensory imaginings have strong referential character toward particulars, such as objects and their property-instances. The explanation of this is that these experiences reveal a substantial portion of the intrinsic characters of some sensory qualities (IR3) and also make the subject aware of the relevant particulars *as* instantiating, or being instances of, those sensory qualities (IR4). That generates strong referential character (IR1).

A final datum is that hard-conscious hallucinations of novel objects, and hard-conscious sensations, do not have strong referential character toward particulars (despite having strong referential character toward sensory qualities, as per the second datum). This can be explained by saying that these experiences do not make the subject aware of any particulars (IR5), so they do not position the subject to make *de re* reference to any particulars. Thus, they lack strong referential character (IR1).

There are some further cases of interest. I have never met Barack Obama, but I have seen him in videos. Evidently, this positions me to make *de re* reference to him. The intrinsic awareness account delivers the further prediction that my experience positions me to know who I am thereby referring to, in the special sense under discussion. After all, as long as the videos are taken in good light, seeing them will reveal to me a substantial portion of the intrinsic character of Obama himself. This prediction seems exactly right. This is interesting: we have found a case in which I know what I am referring to despite having seen only *images* of the target rather than the target itself. The point also extends to types: I have never seen a live zebra, but I have seen videos of them, which will position me to refer to the type *zebra*. According to the intrinsic awareness account, because my experience reveals what tokens of that type are intrinsically like, it will also position me to know what I am thereby referring to. Again, this seems correct.

It is easy to miss the point, since there are many different senses in which you might "know what you are referring to." But once we focus on the sense under discussion, the point comes into sharp focus. If I merely refer to Obama by his functional roles (e.g., as the forty-fourth president of the United States), then I am like the man who refers to the coffee table in his neighbor's home that he has never seen. Once I see a video of Obama, I am like the man after he sees those objects: it is simply revealed to me how the referent is.

Conclusion The intrinsic awareness account has many virtues. For one thing, its core posits are all independently plausible. For another, the account gets the cases right: it correctly predicts that super-blindsight, zombie perceptions, and abstract singular thoughts do not have any strong referential character whatsoever; that all experiences have strong referential character toward some sensory qualities; and that, when they are hard-conscious, all perceptions, all episodic memories, and some sensory imaginings have strong referential character toward particulars. A final virtue is that the account gives us a transparent explanation of what makes experiential knowledge of reference so special: it is that experience reveals a substantial portion of the intrinsic character of the referent.

I take this to be compelling evidence for the intrinsic awareness account. So: which theories of perception can accept the account?

7.4 The Significance of the Intrinsic Awareness Account

Some have thought that only naïve realists can accept anything like the intrinsic awareness account.[8] In this section, I argue that this is false. The account is compatible with *all* of the major theories of perception considered in this book: nonrevelatory/impure revelatory/pluralistic naïve realism, nonrevelatory/impure revelatory/pluralistic representationalism, and rich pluralism.

Most of the commitments of the intrinsic awareness account are obviously compatible with all of these theories. The only commitment that requires any particular scrutiny is this:

IR3. All hard-conscious mental items—that is, experiences—make the subject aware of sensory qualities in a way that reveals a substantial portion of their intrinsic characters.

This claim is perfectly acceptable to impure revelatory naïve realists/representationalists, pluralistic naïve realists/representationalists, and rich pluralists. For all of these theorists accept extended partial revelation—the thesis that, in the first instance, all experiences reveal a substantial portion of the essential truths about the sensory qualities and nothing else. But part of what is essential to any sensory quality is its intrinsic character. Thus, all of these theorists can say that these intrinsic characters are partly revealed in experience.

Nonrevelatory naïve realists/representationalists can also accept IR3. To be sure, these theorists are committed to saying that experiences never reveal the *essences* of anything. Thus, unlike impure revelatory monists and rich pluralists, nonrevelatory theorists will not want to explain IR3 by appealing to anything like extended partial revelation. Still, they are welcome to take IR3, or something like it, as primitive. Nonrevelatory naïve realists can say that hard consciousness is constituted by acquaintance, that acquaintance reveals a substantial portion of the intrinsic characters of its targets, and that all experiences, even hallucinations, acquaint us with sensory qualities. Nonrevelatory representationalists can say the same things, except that they should speak of sensory representation instead of acquaintance.

However, some have argued that representationalists in particular face special problems in explaining strong referential character. Let us appraise these objections.

One objection is this. A hard-conscious perception of a dagger and a matching hallucination differ in their strong referential characters: the perception positions the subject to make knowledgeable reference to a dagger; the hallucination does not. But, the objection goes, the representationalist must say that these experiences have exactly the same strong referential character. After all, the subject can sensorily represent a dagger and its properties whether they are hard-consciously perceiving or hallucinating.[9]

This is a poor objection. According to the intrinsic awareness account, an experience has strong referential character toward a particular dagger only if it positions the subject to make *de re* reference to that dagger (IR1). The representationalist should say that this will not occur merely because the subject deploys a *(putatively) singular representation* of a dagger. The representation must also be *successful*—its content must be constituted by an actual dagger. This occurs in hard-conscious perceptions but not in matching hallucinations. Thus, these experiences will have different strong referential characters.

Another objection is that according to the representationalist, hard-conscious perceptions and singular thoughts are both representations, indeed ones with successful singular contents—contents constituted by real particulars. Thus, hard-conscious perceptions and singular thoughts should be equally effective in providing knowledge of reference of particulars. This is not the case: hard-conscious perceptions have strong referential character toward particulars, and singular thoughts do not.[10]

This objection is also unsound. Given the intrinsic awareness account, the representationalist can say that a mental item has strong referential character only if it reveals a substantial portion of the target's intrinsic character. Hard-conscious perceptions do this; singular thoughts do not. This is a principled way of separating the two cases.

I conclude that the intrinsic awareness account is quite ecumenical: it is available to all of the major theories of perception that we are considering.

To be sure, the account is not *completely* ecumenical. IR3 entails that all experiences reveal a substantial portion of the intrinsic characters of sensory qualities, and as we saw in the last chapter, this will put pressure on *some* views. To begin with, IR3 *might* require us to adopt a primitivist theory of experience: a reductive physicalist theory *might* have trouble permitting genuine revelation of any kind. In addition, IR3 entails that even hard-conscious *hallucinations* reveal a substantial portion of the intrinsic characters of sensory qualities. Again, we saw in the last chapter that this rules out many naïve realist theories of hallucination. However, it does not rule out naïve realism *tout court*. There remain some viable naïve realist theories of hallucination—in particular: **flexible naïve realism**, which treats hallucinations as cases of acquaintance with mind-dependent instances of the sensory qualities; the **imaginative view**, which treats hallucinations as involuntary sensory imaginings of a certain kind; and the **quality-awareness view**, which treats hallucinations as cases of acquaintance with sensory qualities *but not* particulars.

Conclusion The received wisdom is that when it comes to explaining strong referential character, naïve realists have everyone else over a barrel. This is the second of the four core motivations for naïve realism.

I have argued that this assessment is incorrect. The best explanation of strong referential character is the intrinsic awareness account. This account rules out certain naïve realist theories of hallucination and perhaps also reductive physicalist theories. Otherwise, theorists of all kinds are in the game.

7.5 Presentation and Apparent Presentational Character Introduced

I advance now to the third and fourth core motivations for naïve realism, which are interlinked: they have to do with presentation and the appearance thereof. I introduce both motivations in this section. I will scrutinize the third motivation, pertaining to presentation, in the rest of this chapter, while leaving the remaining motivation, which pertains to apparent presentational character, for the next chapter.

Consider these phenomenological descriptions:

> It is a natural, if paradoxical, way of speaking to say that seeing seems to 'bring one into direct *contact* with *remote* objects' and to reveal their shapes and colours. (Broad 1952, 6)

> If, for example, I see a yellow station wagon in front of me, the experience I have is directly of the object. It doesn't just "represent" the object, it provides direct access to it. The experience has a kind of directness, immediacy and involuntariness which is not shared by a belief I might have about the object in its absence. (Searle 1983, 45–46)

> I look out my study window and observe a variegated scene. There are maple, birch, and spruce trees in my front yard. Squirrels scurry across the lawn and up and down the trees. . . . The most intuitively attractive way of characterizing my state of consciousness as I observe all this is to say that it consists of the *presentation* of physical objects to consciousness. . . . It seems for all the world as

> if I enjoy *direct, unmediated* awareness of those objects. There is, apparently, nothing at all 'between' my mind and the objects I am perceiving. They are simply *displayed* to my awareness. (Alston 1999, 182)

> Your visual experience will place a moving rock before the mind in a uniquely vivid way. . . . Visual phenomenology makes it for a subject as if a scene is simply presented. Veridical perception, illusion and hallucination seem to place objects and their features directly before the mind. (Sturgeon 2000, 9)

> The ripe tomato seems immediately present to me in experience. I am not in any way aware of any cognitive distance between me and the scene in front of me. . . . The world is just there. (Levine 2006, 179)

> Something else that we take for granted is that physical objects are the very things that are *presented* to us in perception. It is extremely difficult to make this very natural idea precise. (Brewer 2011, 2)

These quotations, while not saying exactly the same thing, are circling an important phenomenological point. Roughly put, the thought is that hard-conscious veridical perceptions, illusions, and hallucinations have what I will call **apparent presentational character**: they introspectively seem to involve a special kind of sensory awareness, call it *presentation*, that seems to be absent from all other mental items.[11]

We can better lock onto the concept of apparent presentational character by identifying some paradigmatic cases and some foils.

One paradigmatic case occurs when I see a scarlet pepper. Here I introspectively seem to be sensorily aware of the pepper, as well as its instances of scarlet, in a special way: I seem to be *presented* with them. (We can stipulate that presentation seems to be a relation to particulars alone, never to universals.) Another paradigmatic case occurs when I merely hallucinate a scarlet pepper. Again, my experience will introspectively seem to present me with a pepper. More generally, hard-conscious perceptual experiences typically have apparent presentational character.

Apparent presentational character is also found in typical sensations: when I experience an orange afterimage, I introspectively seem to be *presented* with a gauzy strip of orange. To be sure, there introspectively seems to be an important difference between typical perceptual experiences and typical sensations. Ordinarily, a seen or hallucinated maple tree will seem to exist *independently* of my experiencing it. By contrast, in the normal course of things, an orange afterimage will seem to exist *in a way that depends on* my experience of it. But this does not affect the point that, in typical cases, perceptual experiences and sensations both seem to present their targets.[12]

Now consider some foils. First, suppose that after seeing the pepper, I close my eyes and vividly remember, with crisp detail and perfect accuracy, what the pepper looks like. I then form the judgment that the pepper is precisely *that way*. Despite the perfect accuracy of my episodic memory and the knowledgeability of my judgment, neither of these mental items has apparent presentational character: neither seems to give me the right kind of awareness of the pepper. Or suppose that I am a super-blindseer. I have a hard-unconscious perception of the pepper, which causes perfectly accurate beliefs about it to pop into my head in real time. As I walk toward the pepper, a series of beliefs might pop into my head: *there is now a scarlet pepper about 2 meters in front of me—now 1.5 meters—now 1 meter. . . .* In this case, too, I do not seem to be presented with the pepper.

We can pull together all of these ideas to express the phenomenological point. It is this:

> **Apparent presentationality**: Any typical hard-conscious perceptual experience or sensation has *apparent presentational character*—that is, it introspectively seems to *present* one or more particulars. In perceptual experience, these particulars typically seem to exist independently of the subject's mind. In sensation, these particulars typically seem to exist in a way that depends on the subject's mind. However, apparent presentational character is not found in typical hard-unconscious perceptions, typical hard-conscious episodic memories or sensory imaginings, or typical singular thoughts.

It is worth distinguishing apparent presentationality from a well-known and controversial thesis about the "transparency of experience":

> **Transparency**: Take any typical perceptual experience and consider the properties of which the subject is sensorily aware. All of these properties are experienced as properties *of ordinary particulars* that seem to exist independently of the subject's mind; none of them are experienced as properties *of the experience itself*.[13]

Apparent presentationality is logically independent of transparency. On the one hand, transparency is a claim about sensory awareness *in general*. Thus, it does not entail apparent presentationality, which is about a particular *kind* of sensory awareness, namely presentation.[14,15] On the other hand, apparent presentationality does not entail transparency—this, despite the fact that presentation entails sensory awareness. For transparency entails that in typical perceptual experiences, we do *not* seem to be sensorily aware of properties of those experiences. Apparent presentationality does not entail this.

Throughout this book, I am committing only to apparent presentationality. I remain neutral about transparency.

Notice that apparent presentationality only tells us which experiences *introspectively seem* to present particulars. A separate question is, Which experiences *really do* present particulars? It is too early to give a final verdict, but we can reach some initially plausible conclusions.

The easiest cases are hard-unconscious perceptions, hard-conscious episodic memories and sensory imaginings, and singular thoughts. None of these mental items even seem to present their targets. Thus, in the absence of any countervailing evidence, our initial view should be that these mental items do not present anything. It is also easy enough to render an initial verdict about hard-conscious perceptions. These introspectively seem to present mind-independent particulars, and there is no immediate reason to doubt this appearance: the relevant mind-independent particulars are right there in the subject's environment. It is therefore initially plausible to say that hard-conscious perceptions genuinely present what they seem to present.

Now suppose that I hallucinate a scarlet pepper or experience a vivid scarlet afterimage. Might these experiences genuinely present anything? Well, in either case, there will not be a *mind-independent* pepper, or any *external* instances of scarlet, that my experience could present. Thus, if I am presented with any particulars, they must be mind-dependent ones—for instance, sense-data (mind-dependent objects) or mind-dependent property-instances. But the framing assumption from chapter 1 rules out the existence of sense-data. It is also difficult to countenance mind-dependent property-instances, although we have come across one view, *flexible naïve realism*, that posits such things. (I will complete my evaluation of flexible naïve realism in chapter 9.) The most initially plausible view, then, is that hard-conscious hallucinations and sensations do not really present anything.

Putting these ideas together, our default view should be that presentation occurs in all and only hard-conscious perceptions and that, typically, the presented items include ordinary, mind-independent objects or events, along with their property-instances.

There is one last desideratum about presentation to get on the table. It seems obvious that the relation of presentation has a distinctive epistemic value: it seems far better to *see* a scarlet pepper than to have a stream of detailed, accurate singular thoughts about it. Here is one apt expression of the point:

> Suppose that you had a choice between having veridical experiences of your environment, and having a trustworthy, omniscient creature tell you about what's going on in it. I, for one, would prefer to have the veridical experiences. That's not because veridical experiences are the only route to knowledge about my environment—on the face of it, trustworthy and reliable testimony could do just as well. And arguably, the preference isn't just due to the fact that testimony happens to be a slower means of conveying information than veridical experience is—we could revise the case by making the omniscient creature talk really fast and idealizing the subject so that she could understand. I would still prefer to have the veridical experience. (Logue 2012, 227)

The desideratum is to explain what this distinctive epistemic value consists in.

Conclusion We need to account for two important phenomena. The first phenomenon is *apparent presentational character*: for instance, we want to explain why apparent presentationality is true. I will offer my account of this in the next chapter. The second phenomenon is *presentation*: we want to explain why this occurs in all and only hard-conscious perceptions and why it has a distinctive epistemic value. This is the topic of the rest of this chapter.

7.6 Naïve Realism and the Intrinsic Awareness Account of Presentation

Naïve realists often claim to have the best account of presentation: that is the third core motivation for naïve realism. However, the details of the naïve realist's account have not been properly worked out. In this section, I do the ironing on the naïve realist's behalf. I then argue that this account—which I grant is the best account of presentation—is compatible with *all* major theories of perception (sections 7.7–7.8).

We can more explicitly reconstruct the basic motivation for naïve realism as follows. Begin with a phenomenological observation ("PAN" stands for *presentational argument for naïve realism*):

PAN1. Hard-conscious perceptions introspectively seem to present mind-independent particulars.

Now, we should not needlessly fall into an error theory about the introspective appearances. Thus, our default view should be this:

PAN2. If hard-conscious perceptions introspectively seem to present mind-independent particulars, then hard-conscious perceptions really do present mind-independent particulars.

Many naïve realists have then thought that:

PAN3. If hard-conscious perceptions really do present mind-independent particulars, then naïve realism has a major explanatory advantage over all other theories.

The idea is that only naïve realists can do justice to how the presentation relation introspectively seems to be. But these premises entail:

PAN4. Naïve realism has a major explanatory advantage over all other theories.

This is an extremely popular argument.[16]

I accept PAN1 and PAN2. As I see things, the weight of the argument falls on PAN3. What account of presentation might the naïve realist offer to substantiate this premise?

Recall that *naïve realism* is the view that (i) to perceive is to deploy at least one kind of sensory awareness, typically of external particulars and their properties, and (ii) every kind of sensory awareness constitutive of perception is nonrepresentational (i.e., immune to error) in the sense identified in chapter 5. We have seen that the naïve realist might also be attracted to this claim:

> *Acquaintance reveals intrinsic character (ARIC)*: Every acquaintance relation constitutive of hard-conscious perception reveals a substantial portion of the intrinsic character of its targets.

Some have claimed that, when taken against the backdrop of naïve realism, ARIC constitutes the best account of presentation. Their proposal is this ("IP" stands for "intrinsicality" and "presentation"):

The *intrinsic awareness account of presentation (first pass)*:

IP1**. Presentation occurs in all and only hard-conscious perceptions.

IP2**. The distinctive epistemic value of presentation is that it is the only kind of direct awareness that reveals a substantial portion of the intrinsic characters of its targets.[17]

The idea behind IP1** is that presentation is simply acquaintance, which, the naïve realist might say, occurs in all and only hard-conscious perceptions. The idea behind IP2** is to explain the distinctive epistemic value of presentation by appealing to ARIC, the claim that every acquaintance relation constitutive of hard-conscious perception reveals a substantial portion of the intrinsic character of its targets.[18]

I find this account problematic. IP1** and IP2** entail that *only hard-conscious perceptions* reveal any portion of the intrinsic characters of things. But this is difficult to accept: it is plausible that *all* experiences at least reveal a substantial portion of the intrinsic characters of the sensory qualities. Hallucinating scarlet, or vividly imagining it, seems to reveal a substantial portion of the intrinsic character of scarlet just as well as actually *seeing* scarlet.

In response, the naïve realist might try to give as little ground as possible. Regarding hallucinations, they might simply deny that these actually reveal any portion of the intrinsic characters of the sensory qualities. Perhaps they merely *seem* to do so. As for sensory imaginings, perhaps their revelatory power is parasitic on the revelatory power of perception: perhaps a sensory imagining can reveal a substantial portion of the intrinsic characters of sensory qualities only if, and because, the subject has previously perceived those sensory qualities. In this case, the naïve realist can modify IP2**: they can say that the distinctive epistemic value of presentation is that it is a kind of direct awareness that reveals a substantial portion of the intrinsic characters of its targets *in a way that is not parasitic on the revelatory power of any other mental item.*

As I argued in the previous chapter, both of these responses come with significant costs. First, it is just very intuitively plausible that hallucinations reveal a substantial portion of the intrinsic characters of sensory qualities. In response, many naïve realists have provided an error theory to explain away this intuition but that, in itself, is a substantial theoretical cost. Second, it seems possible for imaginative experiences to reveal a substantial portion of the intrinsic characters of sensory qualities in a way that is not parasitic on hard-conscious perceptions. Consider a congenitally blind subject who undergoes a cutting-edge neurosurgery that lets her vividly *imagine* colors despite never having seen them. Afterward, she has her very first color experience: a prolonged, vivid imagining of scarlet. It remains plausible that her imaginative experience will reveal a substantial portion of the intrinsic character of scarlet.

Let us try to develop a naïve realist theory that does not just absorb these costs.

One option for the naïve realist is to grant that sensory imaginings and hallucinations reveal a substantial portion of the intrinsic characters of the *sensory qualities*. Still, they might suggest, only hard-conscious perceptions reveal a substantial portion of the intrinsic characters of *particulars*. This yields:

The *intrinsic awareness account of presentation (second pass)*:

IP1*. Presentation occurs in all and only hard-conscious perceptions.

IP2*. The distinctive epistemic value of presentation is that it is the only kind of direct awareness that reveals a substantial portion of the intrinsic characters of *particulars*.

Unfortunately, IP2* is false. When I vividly remember seeing a scarlet pepper, my experience also reveals to me a substantial portion of the pepper's intrinsic character—but I am not *presented* with the pepper. There must be something more that distinguishes hard-conscious perceptions from hard-conscious episodic memories. (This is also an objection to the first-pass version of the intrinsic awareness account.)

This brings us to a final strategy, and to my mind the best one: divide and conquer. First, the naïve realist can identify what is valuable about hard consciousness in general, whether or not the subject is perceiving. Second, they can identify what is valuable about perception in general, whether or not it is hard-conscious. Then they can tell us what is valuable about these features in combination.

Here is how I think the explanation should go. First, the naïve realist should concede that all experiences, including hard-conscious hallucinations, make the subject directly aware of sensory qualities in a way that reveals a substantial portion of their intrinsic characters. The value of this kind of intrinsicality-revealing awareness is obvious: it is easy to see why any experience of scarlet is special, whether it is a sensory imagining, a hallucination, or a perception.

As for what is special about perception, the naïve realist can borrow my account from chapter 5. There I argued that perceptions have three crucial features: (i) they involve direct awareness of particulars, (ii) they update in real time, and (iii) they tell us how the target is located relative to some part of the subject. I also explained the distinctive epistemic value of these features, both individually and together. One aspect of this value, I suggested, is that perceptions let us interact with particulars in an especially fluid, continuous way.

All that remains is for the naïve realist to explain what is special about the *intersection* of hard consciousness and perception. There is an obvious answer: perception by itself lets us interact with particulars in an especially fluid, continuous way, while hard consciousness reveals a substantial portion of the intrinsic characters of those same particulars. Notice that what hard consciousness reveals is somewhat different for objects than for property-instances. If the particular is an object, then hard consciousness reveals a substantial portion of the intrinsic characters of some of the universals that the object *instantiates*; if the particular is a property-instance, then hard consciousness reveals the intrinsic character of the universal that the property-instance *is an instance of*. Regardless, it seems perfectly clear why this combination is distinctively epistemically valuable: there is special value in having direct awareness of a particular that permits fluid, continuous interaction with it, together with revelation of a substantial portion of the intrinsic character of that particular.

To bring these ideas into sharp relief, we can think about what goes wrong if we start with a hard-conscious perception of a scarlet pepper and then subtract one of the crucial features.

We can start by subtracting hard consciousness. Consider a case of super-blindsight: you look at a scarlet pepper, and although you do not experience it, an involuntary thought pops into your head that there is a scarlet pepper 3 meters in front of you. As you move closer to the pepper, the thought updates accordingly: now you think that it is 2 meters away, now 1.5 meters, and so on. Then some differently colored drops of paint fall onto the pepper, and again your thought spontaneously updates to accurately reflect the pepper's current coloration. But these are all just singular thoughts, without any technicolor experience. There is an obvious and important difference between these super-blindsight-induced singular thoughts and a hard-conscious perception. What is it? If the naïve realist accepts the current account, then they can respond that the singular thoughts merely make you aware of the pepper and its sensory qualities. The hard-conscious perception does more: it reveals to you a substantial portion of the intrinsic characters of those qualities.

Now consider cases in which one of the three crucial features of perception is missing. The first feature is direct awareness of particulars. So suppose that the subject is not directly aware of the pepper. Maybe they are hallucinating, or maybe they are only seeing the pepper's shadow. It is perfectly clear why it would be better to be directly aware *of the pepper itself*.

The second feature of perception that we can subtract is real-time updating. Suppose that you see the scarlet pepper. The next day, you close your eyes and vividly recall, with perfect accuracy, how the pepper used to be. Although your experience makes you directly aware of the pepper and its *past* intrinsic character, it does not make you aware of how the pepper is *now*. That is an important loss.

The final feature of perception that we can subtract is awareness of how the pepper is *in relation to some part of the subject*. Imagine having a perception-like awareness of a pepper that does not place the pepper relative to any part of your body. Such an event would give you a god's-eye sense of the spatial contours of the pepper—a view from nowhere, so to speak. It is easy to see why this is inferior to a hard-conscious perception: it would be almost impossible to manipulate the pepper just on the basis of this kind of awareness.

There is one last kind of case to consider: a case in which the perception and the hard-conscious awareness are not connected in the proper way. Return to the case of super-blindsight: suppose that you have a constantly updating stream of singular thoughts about a scarlet pepper and its features but without any technicolor experience of it. But suppose that you use these thoughts to voluntarily imagine, with perfect accuracy, how the pepper looks from moment to moment. Now you are directly aware of the pepper, and you are aware of a substantial portion of its intrinsic character. You are still not *presented* with the pepper; something important is missing. What is that?

The problem is that your experience of scarlet is not *properly connected* to your direct awareness of the pepper. To test this diagnosis, imagine a subject who has been suffering from super-blindsight. Suddenly, she begins to enjoy hard consciousness when she has visual perceptions: when she looks at a pepper, she experiences the colors, shapes, and textures that the pepper actually has. She takes these to be the colors, shapes, and textures *of the pepper*—not through some voluntary act of imagination but in a spontaneous and immediate way. Clearly, this subject *is* presented with the pepper. Thus, presentation requires an appropriate connection between the subject's perception and her hard-conscious awareness.

In short, the naïve realist's best account is this:

The *intrinsic awareness account of presentation (final version)*:

IP1. What it is for a mental item to present a particular is for it to be a hard-conscious perception of that particular—in particular:

(a) a perception of the particular as either instantiating, or being an instance of, some sensory quality,[19] and
(b) an experience (i.e., a hard-conscious awareness) of that sensory quality, with
(c) an appropriate connection between (a) and (b).

IP2. The distinctive epistemic value of presentation is simply a function of the distinctive epistemic value of hard consciousness and perception—in particular:
(a) All and only experiences reveal a substantial portion of the intrinsic characters of the sensory qualities.
(b) All and only perceptions make the subject aware of particulars in a way that is direct, updates in real time, and varies with the target's spatial position relative to some part of the subject.

Suppose that this is indeed the best account of presentation and that it is uniquely available to the naïve realist. That would support PAN3, the premise that if hard-conscious perceptions really do present mind-independent particulars, then naïve realism has a major explanatory advantage over all other theories.

Conclusion We can now appreciate why the intrinsic awareness account, in its final version, is so appealing.

Part of the appeal is that the account's two main posits enjoy an independent plausibility. The first posit, that hard consciousness reveals a substantial portion of the intrinsic characters of the sensory qualities, wears its plausibility on its sleeve, as we saw in the previous chapter. The second posit, about the three features that characterize perception, is plausible for reasons described in chapter 5.

What is more, the intrinsic awareness account gives us a perfectly transparent link between these posits and the distinctive epistemic value of presentation. It is perfectly transparent why hard consciousness, as well as the three defining features of perception, are all valuable in their own right. It is also perfectly transparent why hard consciousness and perception are valuable in combination. One mark of this value is that only hard-conscious perceptions permit the subject to interact with particulars in a fluid and continuous way while understanding a substantial portion of the intrinsic characters of those particulars.

At the same time, the account does not commit us to implausible claims about experiences that are not perceptions. The account allows that all experiences, including sensory imaginings and hallucinations, can reveal a substantial portion of the intrinsic characters of the sensory qualities and that episodic memories can reveal a substantial portion of the intrinsic characters of particulars. Yet the account also correctly predicts that these experiences do not present their targets.

Thus, I find the intrinsic awareness account to be deeply satisfying. With it in hand, the naïve realist can press the following argument for their view:

The *presentational argument for naïve realism*:

PAN1. Hard-conscious perceptions introspectively seem to present mind-independent particulars.

PAN2. If hard-conscious perceptions introspectively seem to present mind-independent particulars, then they really do present mind-independent particulars.

PAN3. If hard-conscious perceptions really do present mind-independent particulars, then naïve realism has a major explanatory advantage over all other theories.

PAN4. Naïve realism has a major explanatory advantage over all other theories.

Is this argument sound?

7.7 Transferring the Intrinsic Awareness Account: Representationalism and Rich Pluralism

I think the argument fails: I reject PAN3. I agree that the intrinsic awareness account is *so* satisfying that we should all accept it if we can. But I will argue that there is no obstacle blocking rich pluralists, or representationalists of any kind, from doing this.

Here is the intrinsic awareness account of presentation again, in its final form:

IP1. What it is for a mental item to present a particular is for it to be a hard-conscious perception of that particular—in particular:

(a) a perception of the particular as either instantiating, or being an instance of, some sensory quality, and

(b) an experience (i.e., a hard-conscious awareness) of that sensory quality, with

(c) an appropriate connection between (a) and (b).

IP2. The distinctive epistemic value of presentation is simply a function of the distinctive epistemic value of hard consciousness and perception—in particular:

(a) All and only experiences reveal a substantial portion of the intrinsic characters of sensory qualities.

(b) All and only perceptions make the subject aware of particulars in a way that is direct, updates in real time, and varies with the target's spatial position relative to some part of the subject.

Is there anything here that the rich pluralist or the representationalist cannot accept?

It is perfectly obvious that both theorists can accept the entirety of IP1. They will have their own accounts of perception in terms of sensory representations of a particular kind. They will also have their own accounts of hard consciousness. The rich pluralist will explain this in terms of deep awareness; the representationalist might explain it by appealing to a particular functional role, or to higher-order thoughts or perceptions, or they might take hard consciousness as primitive.[20] But none of this makes it even prima facie difficult to accept IP1.

It is also clear that rich pluralists and representationalists can accept IP2. I argued in the last chapter that rich pluralists and representationalists can happily accept part (a)—it is just that they might need to take revelatory awareness as primitive, rather than offering a physicalist reduction of it. Nor does part (b) of IP2 pose any obstacles. This is taken directly from the rich pluralist's account of perception, but the representationalist can adopt it, as well.

Conclusion The intrinsic awareness account of presentation is not uniquely available to naïve realists. Rich pluralists and representationalists can help themselves to exactly the same account. It might be necessary to take it as primitive that experience reveals a substantial portion of the intrinsic characters of the sensory qualities, but this constraint, if it applies at all, would apply to theorists of all kinds.

So far, then, when it comes to accounting for presentation, naïve realists seem to have no particular advantage. But perhaps there is something that we have missed.

7.8 Further Worries about Using Representations to Explain Presentation

Some naïve realists have claimed that if we explain hard-conscious perception by appealing to representations—as representationalists and rich pluralists both do—then we encounter special roadblocks. In this section, I examine these concerns.

One concern is that when I am presented with a scarlet pepper, *the pepper itself* is a constituent of my experience: this much is made plausible by introspection. However, the concern continues, a representationalist or rich pluralist cannot say this. They must say that although a perception and a matching hallucination will have different *causes*, their *constituents* are precisely the same.[21]

This concerns misfires against the representationalist and rich pluralist accounts under consideration. These theorists should grant that a perception of a scarlet pepper and a matching hallucination are both sensory representations and that they are differently caused. In particular, in the perception, the sensory representation stands in a perceptual relation, which is a type of causal relation, to an actual pepper. In a matching hallucination, there is a similar sensory representation, but the perceptual relation is missing. However, the representationalist and rich pluralist are welcome to say that this causal difference *grounds* a constitutive difference, namely, that a perception of a scarlet pepper is a relation of sensory awareness to the particular pepper and its particular instances of scarlet.[22]

It is independently plausible that causal differences can ground constitutive differences. When a wheel of a bicycle stands in certain causal relations to the frame, in particular when the one is *attached* to the other, the wheel becomes a constituent of the bicycle. But suppose you sever those causal relations by detaching the wheel. Even if the wheel is left touching the frame, you have brought it about that the wheel is no longer a constituent of the bicycle.

A related worry is that introspection teaches us that there is an *intrinsic* difference between a perception and a matching hallucination—but representationalists and rich pluralists cannot acknowledge this.[23]

I reply that these theorists can countenance an intrinsic difference if they wish. They can say that when a subject perceives a scarlet pepper, the pepper itself is literally a part of the perception and is thus intrinsic to it. This is so in virtue of there being a certain causal relation between the pepper and certain neural events in the subject. Again, this is analogous to the way in which a wheel can become part of a bicycle in virtue of standing in certain causal relations to it.

I conclude that these objections fail to reveal any advantage for naïve realism over representationalism and rich pluralism.

Still, our work is not quite done: the naïve realist might try to play these same concerns in a different key. They might observe that representationalists say that any perception and matching hallucination have the same *phenomenal character*. However, they might continue, this is not right: when I see a scarlet pepper, the phenomenal character of my experience is partly constituted by, and has as an intrinsic part, that very pepper. The same cannot be true of a matching hallucination. Thus, representationalism is false, and perhaps similar concerns would fell rich pluralism.[24]

I will respond to these concerns after giving my theory of phenomenal character in chapter 10. In the meantime, representationalists and rich pluralists have been given no cause to worry. Like the naïve realist, they can accept that hard-conscious perceptions and matching hallucinations differ in their constituents and in their intrinsic parts.

7.9 Conclusion

It has been thought that naïve realism best explains why hard-conscious perceptions have strong referential character and why they involve the distinctively epistemically valuable relation of presentation: these are, respectively, the second and third core motivations for naïve realism. This chapter has swept these motivations away.

When it comes to strong referential character, I have argued that naïve realists have missed much of the data. Not just hard-conscious perceptions, but *all* experiences, have strong referential character toward the sensory qualities. This occurs because all experiences reveal a substantial portion of the intrinsic characters of the sensory qualities. Meanwhile, if they are hard-conscious, then perceptions, episodic memories, and some sensory imaginings also have strong referential character toward particulars. This occurs because these experiences have additional revelatory power concerning the intrinsic characters of particulars.

Then there is presentation. I argued that this occurs only when there is hard consciousness, for this is what reveals a substantial portion of the intrinsic characters of things. I also argued that presentation occurs only in perceptions, for only these make the subject aware of particulars in a way that is direct, updates in real time, and varies with the target's spatial position relative to some part of the subject. The distinctive epistemic value of presentation is a function of the distinctive epistemic values of hard consciousness and perception.

These accounts of strong referential character and presentation are available to naïve realists. But *exactly the same accounts* are available to representationalists and rich pluralists. Thus, no one has any comparative advantage here.

8 Apparent Presentational Character

Out of the four original motivations for naïve realism, only one remains: the idea that it offers the best account of apparent presentational character. In this chapter, I dismantle this motivation. More generally, I offer a theory of when apparent presentational character occurs and what its nature is.

I begin by briefly reviewing the earlier data about apparent presentational character (section 8.1). Then I consider an argument that naïve realists have the best account of apparent presentational character, and I show that this argument fails. I go on to examine some tempting but unsuccessful accounts of apparent presentational character (section 8.2). Next, I lay out my own account of apparent presentational character, the **categorization account** (section 8.3). After demonstrating its initial explanatory power (section 8.4), I rebut some objections (sections 8.5–8.6) and discuss the epistemology of our beliefs about presentation (section 8.7). I then compare the categorization account to two more rival accounts (sections 8.8–8.9). I round things out by considering how the categorization account bears on the philosophy of perception more broadly (sections 8.10–8.11).

8.1 Apparent Presentational Character Revisited

The previous chapter offered an account of *(actual) presentation*—an account of why certain experiences *in fact* just present their subjects with various particulars. But we also need an account of *apparent presentational character*—an account of why certain experiences *introspectively seem* to present particulars and why those particulars *introspectively seem* to be mind independent or mind dependent.

Recall that we need to explain the following:

> *Apparent presentationality*: Any typical hard-conscious perceptual experience or sensation has *apparent presentational character*—that is, it introspectively seems to present one or more particulars. In perceptual experience, these particulars typically seem to exist independently of the subject's mind. In sensation, these particulars typically seem to exist in a way that depends on the subject's mind. However, apparent presentational character is not found in typical hard-unconscious perceptions, typical hard-conscious episodic memories or sensory imaginings, or typical singular thoughts.

Notice that apparent presentationality speaks only of what is *typical*. As I argued in chapter 5, some atypical sensory imaginings and episodic memories *do* have apparent presentational character—for instance, the flashbacks of a war veteran.

Bearing this in mind, how might we account for apparent presentational character?

8.2 Some Initially Tempting Accounts

In this section, I consider a series of initially tempting accounts of apparent presentational character. I argue that none of them are adequate.

One initially tempting thought is that any genuinely presentational experience wears its presentational status on its sleeve. That is:

> The **easy account**: Take any subject whose experience presents them with a mind-independent/mind-dependent target and who possesses the concepts of presentation and mind independence/mind dependence. Just on the basis of careful, rational reflection on their experience, it will introspectively seem to them that they are presented with a mind-independent/mind-dependent target.[1]

Perhaps the easy account could be extended to sensory imaginings and episodic memories: perhaps these wear their nonpresentational status on their sleeves, too. We would still need to explain why hallucinations and sensations incorrectly seem to present mind-independent and mind-dependent targets, respectively. But the account gives us someplace to start.

What is more, the easy account is an essential piece of the fourth and final motivation for naïve realism. The idea is that the easy account is the best account of apparent presentational character, insofar as it is very simple and does full justice to the introspective appearances. But this account is only available to naïve realists: it relies on the claim that perceptions *really do* present their targets, and only naïve realists can fully honor that idea.

I offer two responses to these ideas.

First, even if the easy account were true, it would not support naïve realism. As I argued in the last chapter, the best account of presentation is the intrinsic awareness account, which is not uniquely available to naïve realists. It is equally available to representationalists and rich pluralists. Thus, they can also accept the easy account.

A further problem, however, is that the easy account is incorrect. Suppose that, unbeknownst to me, you set up faint flashing lights within my field of vision. When I see them, my experience is indiscriminable from the phosphene experience that I get when I vigorously rub my eyes. The easy account predicts that it will introspectively seem to me that I am presented with mind-*independent* lights. That is incorrect: it will introspectively seem to me that I am presented with mind-*dependent* phosphenes. To make this perfectly clear, suppose that I vigorously rub my eyes just before I see the faint flashing lights. I simultaneously experience phosphenes and real lights, yet I might well be unable to tell the two apart.

Cases like this are not hypothetical. They are actual. In order to test peripheral vision, ophthalmologists position patients in front of a half-sphere that produces flashing lights, some of which are extremely faint. I have done this test a few times, and unlike the subject described in the previous paragraph, I knew the exact setup beforehand. Still, I often mistook phosphenes for real lights and vice versa.

It is easy to multiply examples of this sort: many sensations are introspectively indiscriminable from matching perceptions. A vivid sensation of a bright green afterimage might perfectly match a perception of a green light; a sensation of bitterness on the tongue might perfectly match a perception of a faintly bitter taste; a sensation of ringing in the ears might perfectly match an actual faint ringing sound. Again, this last kind of case is actual, not hypothetical. It occurs in hearing tests. Under elaborate experimental conditions, it is even possible to create vivid, detailed afterimages in which subjects introspectively take themselves to be perceiving *entire scenes* in their full visual glory. Here is how one scientist describes his own experience during such an experiment:

> I could scarcely convince myself that the light was not still burning; it seemed as if I could see my real hand, the real objects on my table and even the pictures on the wall. These decidedly positive afterimages persisted for many seconds. . . . When I fixate with both eyes and then close them carefully after the last illumination, it is often difficult for me to believe that they are really closed. Observers often exclaim, 'My eyes won't shut,' 'I've lost control of my eyelids,' etc. This illusion is evidently due to the fact that the after-images, which have the same appearance, as far as colour and form are concerned, as the illuminated objects, are not only just as distinct as when the eyes are open but usually much more so. (Swindle 1916, 329)

The lesson is that some genuinely presentational experiences do not wear their presentational status on their sleeves. Some perceptions present mind-*independent* targets that introspectively seem to be mind *dependent*. And perhaps some sensations present mind-*dependent* targets (although I seriously doubt it, for reasons discussed earlier), but those targets introspectively seem to be mind *independent*.

This might inspire us to weaken the easy account. Even if not all perceptions wear their status as *presenting mind-independent/mind-dependent particulars* on their sleeves, they might still wear on their sleeves that they *present* something, full stop. The idea is this:

> The **weakened easy account**: Take any subject whose experience presents them with something and who possesses the concept of presentation. Just on the basis of careful, rational reflection on their experience, it will introspectively seem to them that they are presented with something.

The weakened easy account is even less complete than the easy account, since it does not even tell us why apparently presented particulars seem to be mind independent or mind dependent. But is the account at least correct?

It is not: all hard-conscious perceptions present their targets, but sometimes they introspectively seem not to do so. In one famous experiment, subjects looked at a white wall and were asked to imagine a blue banana rotating slowly on the wall. Unbeknownst to them, the experimenters then projected a dim image of a rotating blue banana on the wall. Subjects did not realize that they were seeing a projection of a banana. They believed that they were merely sensorily imagining the banana.[2]

Some will worry that a subject of this experiment *can* tell that their blue banana experiences are perceptions, since these experiences are involuntary. Suppose that this is the correct verdict about this particular case. Still, I think the verdict cannot be correct in general: it is possible for an involuntary experience to introspectively seem to be voluntary.

To see this, consider once more a subject who is instructed to look at a white wall and imagine a blue banana on it. This time, however, the subject is told that scientists will sometimes be projecting a blue banana image on the wall. Their task is to determine whether each of their blue banana experiences is a perception or a sensory imagining. Suppose that the experiment is run by future neuroscientists who can detect, through advanced neuroimaging, exactly when the subject forms the intention to imagine a blue banana. These scientists can predict the exact shape, orientation, color, and texture that the imagined banana would have. Whenever the subject is about to imagine a blue banana, the scientists project a faint image of a banana on the wall that has those exact qualities. The projected image will even rotate, change color, and so on in accordance with the subject's intentions.

A subject of this experiment would mistakenly take their perceptions to be voluntary. These perceptions would not introspectively seem to present anything, even though they would in fact be presenting blue banana images. This restores the counterexample to both versions of the easy account.

Such a counterexample can even be developed without any science-fictional elements. (Serendipity is required instead.) Imagine a subject in a grassy field who is about to imagine a firefly in their surroundings. They intend to imagine the firefly traveling in a very specific pattern. By chance, they see a real but very faintly lit firefly exactly where they had intended to imagine one, and by even greater chance, the firefly follows the precise path that they had intended to imagine it following. This subject is really presented with a mind-independent firefly, but it will introspectively seem to them that they are not presented with anything.

The moral is that neither version of the easy account will work.

These accounts have a further weakness. We want to know not just when apparent presentational character *occurs* but what apparent presentational character *is*, metaphysically speaking. Neither version of the easy account answers this question.

Let us consider some further options. Once we notice that there is some correlation between experiences with apparent presentational character and experiences that are involuntary, we might land on this account:

> *Involuntarism*: For an experience to appear to present a particular is for it to introspectively seem to involve involuntary sensory awareness of that particular.

This account is no good. If you tell me *not* to think about tomatoes, I might end up vividly picturing one anyway, despite my best efforts to the contrary. My sensory imagining will be involuntary in a way that is introspectively evident, so the involuntarist predicts that my experience will have apparent presentational character. This is not the case.

But observe that apparent presentational character is also correlated with vivid, determinate experiences. That suggests the following proposal:

> The **determinate content account**: What it is for a mental event to introspectively seem to present a *mind-independent* particular is for it to be a sensory representation *with a determinate content.*

This account is false. During sparring sessions, I have, regrettably, been struck in the head hard enough to experience vivid flashes of light. My experiences had highly determinate contents, but they did not for a moment strike me as presenting mind-independent flashes of light. It was evident to me that these were sensations.

Here is a somewhat better account in the same spirit:

> The **modified determinate content account**: What it is for a mental event to introspectively seem to present a particular, *whether mind independent or mind dependent,* is for it to be a sensory representation with a determinate content.[3]

This account gets the sparring case right: my experience of a bright flash has a determinate content, and it introspectively seems to present a particular, albeit mind-dependent, light. Still, the account fails in both directions. In some respects, it is too liberal. It is certainly metaphysically possible to have sensory imaginings that are extremely vivid. This is probably actual: some chess players report being able to visualize a chessboard in quasi-perceptual detail. But such experiences will not seem to be *presentational.* In other respects, the modified determinate content account is too strict. Suppose that you close your eyes and then open them as little as possible to see a cup in front of you. You will barely be able to make out the cup, and your experience will be very faint. Still, you will introspectively seem to be presented with a mind-independent cup.

This last case is also a counterexample to the following account, which appeals to involuntariness *and* determinate content:

> The **involuntary determinate content account**: What it is for a mental event to introspectively seem to present a particular, whether mind

> independent or mind dependent, is for it to be an *involuntary* sensory representation *with a determinate content.*

A faint visual experience of the cup does not satisfy this condition, but ordinarily, it will still seem to present a mind-independent cup.

Conclusion Easy accounts are false: experiences can present their targets without introspectively seeming to do so. Nor can apparent presentational character be explained by a straightforward appeal to the involuntariness and/or determinacy of an experience. What else might we try?

8.3 The Categorization Account Introduced

In my view, the best account is this:

> The *categorization account*: For an experience to introspectively seem to present a target, and for the target to introspectively seem to be mind independent or mind dependent, is for the subject to *categorize* the experience or its targets as such. In normal, mature human beings, this categorization occurs on the basis of an overall gestalt.[4]

This account requires much elaboration.

First elaboration What is *categorization*? It is the process of applying concepts automatically and subpersonally, in a way that is not fully sensitive to the subject's current background beliefs. I leave open the possibility that categorization can occur without any connection to experience. Still, it is paradigmatically associated with perception: when you see something that looks to have certain distinctive pattern of shapes and colors, you categorize it *as* a maple tree. You would categorize it in this way even if you believed it to be a movie prop made of cardboard. Your contrary background beliefs would not stop the categorization from occurring.

I will suggest in section 8.5 that the categorization described in this account—the categorization that constitutes apparent presentational character—occurs within experience proper.

Second elaboration What is it for an experience to be categorized as *presenting* its target? The idea is that as we mature, we begin to develop a *tacit partial theory* of what is occurring in experiences of various kinds. At the same time, we use that theory to develop a *concept* of presentation, which can be applied via categorization.[5]

In particular, over the course of our experiences when we are very young, we notice that some of our experiences are voluntary—to a first approximation, their specific characters are sensitive to our will. We can choose to imagine the scent of vanilla or to remember the tune of "Paint It, Black." However, we also observe that some experiences—perceptions and sensations, for instance—are not like that. We have no volitional control over their specific characters. What could determine the character of our experiences in these cases, then? We tacitly reason that it is the character of certain target(s) that we are aware of, targets that exist concurrently with the experiences themselves.

In this way, we form the concept of *presentation*, albeit without naming it that. This is the concept of whatever best fits platitudes such as the following (where *P* stands for both *platitude* and *presentation*):

P1. Such-and-such experiences (which are in fact perceptions and sensations) present their targets.

P2. Presentation entails involuntary experience.

P3. The character of a presentational experience is largely determined by the character of the concurrently existing target(s) of that experience. Thus, how the character of a presentational experience unfolds over time is largely determined by how the target itself unfolds over time.

P4. Presentation is a natural kind.

We then automatically and subpersonally apply this concept of presentation, via categorization, to certain experiences, including typical perceptions, hallucinations, and sensations.[6]

Focus for a moment on the first half of P3, the claim that the character of a presentational experience is largely determined by the character of the concurrently existing target(s) of that experience. This claim is false: there is a time lag required for perceptual processing. The lag is ordinarily very short, but in principle, it can be arbitrarily long: think of what happens when we see stars. Does this show that our concept of presentation does not refer? Of course not! The concept refers to whatever fits the platitudes *best* (as long as the fit is reasonably good). A perfect fit is not required.

Third elaboration What is it for an experience to be categorized as presentational, and for its targets to be categorized as mind independent or mind dependent, *based on an overall gestalt*? This question brings us to the heart of the categorization account.

To illustrate the idea, consider a mundane case in which you see a tomato. There are many cues that you are *presented* with a tomato rather than sensorily imagining or episodically remembering it. There are also many cues that the tomato, unlike an afterimage or a phosphene, is *mind independent*.

Some of these cues primarily involve the sensory qualities *of the tomato*: the redness of the tomato is experienced as especially vivid, as relatively stable, and as part of a continuous three-dimensional surface. Then there are other sensory cues that involve the *interaction* between the sensory qualities of the tomato and the sensory qualities of your environment or body. For instance, you may lose your awareness of the redness of the tomato as the lights dim. In addition, your awareness of redness will persist only while you are looking in the right direction with your eyes open, and there will be changes in the shape that you experience as you move around the tomato. One last important factor is that your experience of these qualities will seem to be *involuntary*. Together, these cues constitute what I will call a *perceptual gestalt*—a gestalt taken to indicate that you are being presented with a mind-independent target.

For another illustration, suppose that you experience an orange afterimage. Again, the categorization account says that you rely on a gestalt of cues to categorize your experience. As before, your experience is involuntary and vivid. Thus, you categorize the experience as presentational (incorrectly, according to the rich pluralist anyway—they hold that sensations do not really present anything). However, the afterimage does not seem to be well integrated with the environment, nor does your awareness of it seem to change in predictable ways as the lights dim or as you move or close your eyes. You eventually learn that others do not seem to be aware of your afterimages, either. For these reasons, you categorize the afterimage as mind dependent. In fact, once you learn what is characteristic of an afterimage, you will categorize sensorily similar targets as mind dependent *even before* you have tried moving around or closing your eyes. This time, your experience has a *sensational gestalt*—a gestalt taken (perhaps mistakenly) to indicate that you are being presented with a mind-dependent target. Experiences with a perceptual or sensational gestalt will share a more general *presentational gestalt*: a gestalt taken to indicate that you are being presented with *something*, whether mind independent or mind dependent.

According to the categorization account, no single (nonweighted) set of cues is necessary and sufficient for categorizing an experience as perceptual, sensational, or presentational. Instead, each of these categorizations occurs when and because you identify a gestalt of cues that you have learned to recognize on the basis of much previous experience. *That* is the relevance of features like involuntariness and determinacy: these are cues with heavy but nonmaximal weight.

That is the categorization account. Why accept it?

8.4 Evidence for the Categorization Account

I submit that the categorization account explains all of the data so far, as well as some new data.

Datum Some hard-conscious perceptions—which always genuinely present mind-independent targets—may introspectively seem to present a mind-dependent target: recall the subject who rubs their eyes and sees faint, mind-independent lights but takes themselves to be aware of mind-dependent phosphenes. Other perceptions do not introspectively seem to present anything at all. Recall the subject who thinks that they are imagining a blue banana rotating slowly on a wall but actually sees a faint projection of a rotating blue banana, as well as the subject who thinks they are imagining a firefly tracing a particular path in a field but are actually seeing such a creature.

The categorization account explains all of this. To begin with, it allows that a perception may sometimes lack a perceptual gestalt. This is what happens to the subject who sees faint flashing lights. Their experience is involuntary, so they correctly categorize it as presentational. But their perceptions of those lights are not vivid, given that the lights are faint; their perceptions are not stable, given that the lights flash only briefly; and their perceptions are not well integrated with the environment, since the lights are well hidden. They have also just rubbed their eyes, which, as they have previously learned, is usually followed by an experience of what they take to be mind-dependent lights. In short, their perception has a *sensational* gestalt, which causes them to incorrectly categorize the targets of their experience as mind dependent.

The categorization account allows that a perception may not even have a *presentational* gestalt. That is what happens in the blue banana experiments. Here the subject's perceptions are not at all vivid, and the targets of their perceptions are unusual: we do not usually come across faint projections on walls. Most importantly, these experiences introspectively seem to be voluntary, for just when the subject forms an intention to visualize a blue banana on the wall, they have such an experience. Thus, it introspectively seems to them that they are not introspectively presented with anything at all. The same kind of story can be told for the subject who sees a firefly but thinks that they are imagining it.

A pair of data Some highly determinate experiences seem to present mind-*dependent* targets, as when I am struck in the head and experience bright flashes of light. In addition, some highly *in*determinate experiences *do* seem to present mind-independent targets, as when you close your eyes and then open them as little as possible to see a cup in front of you.

The categorization account predicts these results, since it does not appeal to any single experiential cue to explain apparent presentational character, but to gestalts, which are holistic. When I am struck in the head, before I experience vivid flashes, I feel the sensory quality of a sharp blow to my head, which I have learned is often followed by an experience of a vivid flash. I have also found no reason to think that getting hit creates real, mind-independent flashes in my environment. Thus, my experience of the flash does not seem to present me with anything mind independent. Similarly, when you faintly see the cup, you can feel the sensory qualities of your eyelids that indicate that your eyes are not fully open, so you are not surprised to find that your visual experience is faint—not to mention that your experience of the sensory qualities of all other seen objects is faint, too. On top of that, your experience of the cup has many other sensory similarities to a normal perception of a cup: for instance, you experience the sensory qualities indicative of a table supporting the cup. That is why this experience will strike you as presenting a mind-independent cup.

Datum It is possible for the introspective appearances regarding presentation to conflict with what you know. You might know that you are about to see faint lights that look just like afterimages or phosphenes. Still, if the timing of the lights is precise enough—if they go off just when you rub your eyes and fade shortly afterward—then it will seem to you that you are not presented with mind-independent lights. Or you might know that your brain is about to be stimulated to create a perfect hallucination of a maple tree. The hallucination, when it occurs, will still introspectively strike you as presenting a mind-independent tree.

The categorization account explains this by saying that categorization occurs *automatically and subpersonally*. It is not sensitive to the entirety of the subject's background knowledge at a given moment, just as the Müller–Lyer illusion does not end simply because the subject knows that it is an illusion. This is compatible with categorization resulting from perceptual learning, since such learning occurs over extended periods of time.

This also makes it clear how we should handle an important phenomenological objection. The objection is that according to the categorization account, the introspective appearance of presentation is phenomenologically indirect: you identify your experience as having a certain gestalt, and from that you arrive, whether by inference or by some other transition, at the introspective appearance of presentation. But this is plainly incorrect: the introspective appearance is phenomenologically direct. It just immediately seems to you that you are presented with the tomato.

I agree with my opponent that the introspective appearance is phenomenologically direct. That is precisely what the categorization account says: it is part of the view that your categorization of the relevant gestalts occurs *automatically and subpersonally*. The process takes no noticeable effort and is not within the realm of introspective access. Thus, the introspective appearance of presentation just shows up in your phenomenology.

Datum There is an interesting empirical phenomenon known as *derealization disorder*. It occurs in certain pathological subjects, as well as in ordinary subjects who undergo hypnosis or sensory deprivation. These subjects have no obvious perceptual deficits—they can perceptually track and identify objects just as well as ordinary subjects—but they report that their experiences feel unreal, as though they are dreaming or watching a movie.[7]

The categorization theorist can easily explain what is happening in such subjects: their mechanisms for categorization are not working properly. This explains why there is a feeling of unreality: this consists in the absence of certain categorizations that would normally occur, such as categorizations of presentation and mind independence. It also explains why these subjects have no *perceptual* deficits. For the constituents of hard-conscious perception—whether these are sensory representations, acquaintance, or something else—are entirely metaphysically independent of the categorizations that are missing in derealization subjects.

Datum We return now to a familiar observation in the philosophy of perception. Take any perception that introspectively seems to present mind-independent particulars. Any matching hallucination will introspectively seem to do the same.

The categorization account explains this, at least *if* we conjoin it with representationalism or rich pluralism. (I will return later to the question of whether the explanation is available to naïve realists.) For the categorization account says that all introspective appearances of presentation in normal, mature human beings are explained by categorizations based on a perceptual gestalt. This gestalt is found in any typical perception. To this, representationalists and rich pluralists add that for any perception, there is a matching hallucination, which is analyzed as a hallucination that consists of sensory representations of the same kind (modulo whether they have successful or failed singular content). Rich pluralists add that in hard-conscious cases, there is also deep awareness of the same profile of sensory qualities. Regardless, such a hallucination will have exactly the same gestalt as the perception it matches. Thus, any given subject will categorize these experiences in exactly the same way.

Indeed, the categorization account lets us answer a more basic question about hallucinations. I proposed in chapter 5 that to be a hallucination is to be a sensory representation that is *mistakenly* taken to be a perception. This provokes a question: why do hallucinations occur in the first place? That is, why would we ever mistake *other* kinds of sensory representations for perceptions? The categorization account lets us say: in normal, mature human beings, this occurs when and because a sensory representation is not a perception but is still taken to have a perceptual gestalt.

Conclusion Other accounts were unable to explain the full range of data about apparent presentational character. The categorization account does better. But let us make a more thorough examination of it.

8.5 Can Categorization Occur within Experience Proper?

The following claim is introspectively plausible:

APE1. For ordinary human subjects, apparent presentational character sometimes (and arguably always) occurs within experience proper.

("APE" stands for "apparent presentational character is properly experiential.") However, some would argue that:

APE2. For ordinary human subjects, categorizations never occur within experience proper.

For some would say that it is metaphysically possible for categorizations to occur within experience proper; others would say that such categorizations, while metaphysically possible, never occur in ordinary normal human subjects. Thus, by a mild extension of Leibniz's law:

APE3. Apparent presentational character is not constituted by categorizations.

But APE3 is incompatible with the categorization account.

I accept APE1, so I must reject APE2. Justifying this move is the task of this section. First, I argue that it is metaphysically possible for categorization to occur within experience proper. Then I sketch some empirical evidence that this possibility is often realized in us.

On the metaphysical possibility of categorization There is no obvious contradiction in the idea that categorization can occur within experience proper. Thus, I take it that our default assumption should be that this is a genuine metaphysical possibility, barring some good argument to the contrary. My approach, then, will be to rebut some initially plausible arguments against this possibility. I see three constituencies—conservatives about hard character, nonconceptualists, and naïve realists—who might offer such arguments.

Start with **conservatism about hard character**. To understand the defining commitment of this view, consider the recent controversy over what happens when (and if) we perceive high-level features—for example, when we see a tree *as a pine tree* or hear a sentence *as being in Russian*. The controversy is over whether there is any *proprietary hard character* associated with perceiving high-level features. **Liberalism about hard character** is the view that there is. For example, a liberal might say that there is some hard character associated with an arborist's seeing a pine tree that is something over and above the hard character associated with them seeing the pine tree's various shapes, colors, textures, and so on.[8]

Conservatives about hard character dissent. They think that the hard character of perception is exhausted by the hard character associated with the low-level features (and perhaps objects) that we perceive; if there is such a thing as perceiving high-level features, then this results in changes in hard character only insofar as it causes changes in which low-level features (and perhaps objects) we perceive.[9] A conservative about hard character might think that this commits them to holding that it is metaphysically impossible for categorization to occur within perception proper.

This is not so. Conservatism is a view about the *hard character* of perception. It is perfectly compatible with the claim that categorizations can occur within perception proper, *as long as categorizations are not taken to generate any hard problems of consciousness*. This point is well appreciated by hard-character conservatives themselves. For instance, Carruthers and Veillet assert that "[a] property is phenomenal . . . only if it gives rise to an explanatory gap" (2011, 45). Thus, they are using the expression *phenomenal consciousness* with hard consciousness in mind. (I defend this claim further in chapter 10.) They go on to say:

> The initial nonconceptual outputs of early vision are monitored by a whole host of conceptual systems . . . the relevant concept is attached to the nonconceptual representation in question and globally broadcast along with it, for other concept-wielding consumer systems to take note of and draw inferences from. . . . [However, from] the fact that a given concept is a constitutive *component* of a phenomenally conscious [i.e., hard-conscious] perceptual state it doesn't follow that the concept makes a constitutive *contribution* to the phenomenal qualities of that state. (Carruthers and Veillet 2011, 42)

But there is a second constituency, *nonconceptualists* about experience, who might think that they are committed to denying the metaphysical possibility of properly experiential categorization. Nonconceptualists think that experiences essentially have *nonconceptual content*. While there are importantly different ways of spelling out precisely what this means, the basic idea is that a subject can have an experiential state with a particular content <*p*> whether or not that subject has the concepts required to entertain <*p*>.[10]

However, nonconceptualism is just the view that experiences *do* essentially have nonconceptual content; perhaps some will also build it into the view that experiences do *not* essentially have conceptual content. All of this is compatible with the claim that some experiences *accidentally* have conceptual content that is layered on top of its nonconceptual content.[11]

We can trace a similar dialectic for naïve realists. A central commitment of naïve realism is that any perception essentially involves *nonrepresentational* awareness. But categorization is evidently *representational*. So, naïve realists might reason, they should say that it is metaphysically impossible for categorization to occur within perception proper.

This reasoning is fallacious. Yes, naïve realism says that any perception *essentially* involves nonrepresentational awareness, and some would add that perception does *not* essentially involve representation. All of that is compatible with the view that some perceptions *additionally and accidentally* involve representation. This point has been recognized by naïve realists for decades:

> [Naïve realism] simply consists in the insistence that perception essentially involves a mode of cognition of objects that is nonconceptual in character. Moreover it is that mode of cognition that gives perception its distinctive character vis-a-vis other modes of cognition—abstract thought, fantasy, memory, and so on. But this insistence does *not* commit [the naïve realist] to the denial of [the thesis that] . . . perception is typically conceptually structured. . . . Indeed, I accept that . . . when I look out my study window . . . I see various parts of the scene *as* houses, trees, etc., employing the appropriate concepts in doing so. (Alston 1999, 184)

That concludes my argument that it is metaphysically possible for categorization to occur within experience proper. Briefly put, the argument was that this scenario seems to be perfectly coherent, and there is no good reason—not even for hard-character conservatives, nonconceptualists, or naïve realists—to deny that this is metaphysically possible.

Still, is properly experiential categorization *actual*?

On the actuality of categorization I think so. There are many empirical results that support this claim.[12]

In one telling experiment, subjects were presented with a word, such as "flowers," and then shown a series of images. Each image was presented for *just 13 ms*. For comparison, a single blink of the eye takes *at least* 100 ms, so these images were presented for about one-eighth of the time that it takes to rapidly blink your eyes. Moreover, these images were forward- and backward-masked to disrupt any further visual processing. The idea was to ensure that subjects' visual systems really had only 13 ms to process the images. For each image presented in this way, subjects had to say whether or not it contained flowers. If the image did contain flowers *and* if subjects correctly noted this fact, then they were presented with two more images, both of which contained flowers, and they had to identify which of these images they had already seen. Astonishingly, despite these remarkably brief presentation times, subjects were able to perform this task at well above chance.[13]

These subjects were presumably performing genuine *categorization*—that is, they were genuinely applying *concepts*. For they had to apply linguistically presented information (the word "flowers") to what they perceived and use that to guide their actions (pressing the "yes" or "no" buttons). The use of a single representation across such varied mental systems is a hallmark of categorization.[14] Moreover, it seems that this categorization was occurring *experientially* rather than cognitively. For *even within the visual system*, it takes about 50 ms for feedback processing to occur, and feedback processing involving cognition proper would take much longer than that. Given that the presentation times in the current experiment were just 13 ms, there was simply not enough time for this categorization to have been cognitive.

This interpretation is reinforced when we examine overall response times on tasks that can be performed very quickly. For example, in another study, subjects were presented with two images of a natural scene for just 20 ms per image. One image contained an animal in a landscape, while the other contained only a landscape. The subject's task was to saccade to the image containing the animal. Subjects were able to perform this task—which required perceiving, categorizing, and preparing and performing the saccade—in just 120 ms. Again, this is just too little time for the categorization of animals to have occurred cognitively.[15]

This is just a taste of the relevant evidence. There is much more like it.[16]

Conclusion In this section, I have engaged with the following objection:

APE1. For ordinary human subjects, apparent presentational character sometimes (and arguably always) occurs within experience proper.

APE2. For ordinary human subjects, categorizations never occur within experience proper.

APE3. Apparent presentational character is not constituted by categorizations.

I accept APE1, but I have argued that we should reject APE2: all parties should agree that it is metaphysically possible for categorization to occur within experience proper, and the empirical evidence suggests that this kind of categorization is a routine occurrence.

But let us consider some further objections to the categorization account.

8.6 Other Objections and Replies

In this section, I rebut, first, some objections that target the categorization account by itself and, second, some objections that target the account only when it is conjoined with representationalism or rich pluralism.

Objection Grant for the sake of argument that *once a subject already appreciates that certain particulars are mind independent*, they can learn to identify the gestalts associated with perceiving those particulars. Still, you might wonder what gets this process off the ground. How can a subject come to appreciate, as the categorization account requires, that anything is mind independent in the first place?

Reply I believe that there is a great deal that might tip them off to this fact. For instance, they can easily learn through experience that a cup will fall if it is not supported by anything. Suppose, then, that they see a cup supported by a table. They position their hand so that it covers the table entirely while leaving the cup in plain sight. If the table were mind dependent—if it ceased to exist now that the subject was no longer perceiving it—then the cup ought to fall. That does not happen. It is easy for a subject to pile up evidence of this sort.

The objection continued You might still think that this reply presupposes too much: how can the subject form the *concept* of mind dependence or independence in the first place?

Reply Notice that perception does not, by itself, reveal anything about *what it is* to be a mind. Perception does not even make us directly aware of minds: when I see a cup, it seems to me that I am directly aware of the cup but not of my own mind.[17] So I think that we must form the concept of a mind, and relatedly the concepts of mind dependence and independence, simply by positing a kind of entity that plays various theoretical roles. Once that is granted, I cannot find any difficulty, in principle, for the claim that we can form such a concept on the basis of experience.

I turn next to objections that target the categorization account when it is conjoined with representationalism or rich pluralism.

Objection On the representationalist account, perception provides the subject with nothing more than a sensory representation (i.e., a mind-dependent image). But the subject cannot extract a conception of mind independence from something that is mind dependent. Thus, representationalists cannot explain why perceived objects typically introspectively seem to be mind independent.[18]

Reply When the representationalist says that perception *provides* the subject with nothing more than a sensory representation, they mean that part of what it is to perceive is to deploy a sensory representation. They do *not* mean that in perception, the subject is *directly aware of* nothing more than a sensory representation. To begin with, on most representationalist accounts, the subject need not be aware of sensory representations at all.[19]

More importantly, representationalists almost always hold that in perception, the subject is directly aware of—by sensorily representing—*particular mind-independent objects and their property-instances*. So the problem of deriving a conception of mind-*independent* entities from awareness of mind-*dependent* entities does not even arise.

Objection Perceptual representation and categorization are each just further forms of representation, on a par with thought, speech, and signs. Thus:

MI1. If the introspective appearance of objects as mind independent is merely a matter of perceptually representing (or categorizing) those objects as mind independent, then perception is no more fundamental than other forms of representation for appreciating the mind independence of objects.

Yet it seems obviously true that:

MI2. Perception is more fundamental than other forms of representation for appreciating the mind independence of objects.

Thus:

MI3. The introspective appearance of objects as mind independent is not merely a matter of perceptually representing (or categorizing) those objects as mind independent.[20]

Reply My reaction to this argument depends on how the phrase *appreciating the mind independence of an object* is to be understood.

Perhaps the idea is just that seeing a tomato positions the subject to know *which* object they are representing as mind independent. On this interpretation, what is special about perception is the way that it relates the subject to the object, not to the property of mind independence. In that case, I would reject MI1. I have already explained that theorists of all stripes can explain this special feature of perception via the intrinsic awareness account: they can say that the experience has this feature in virtue of revealing part of the very intrinsic character of the sensory qualities of the object. This does not happen in thought.

But there is a bolder interpretation, on which seeing a tomato positions the subject to appreciate the mind independence of objects in a much stronger sense. The idea is that what is presented when you see a tomato is not only the tomato but also its *mind independence*. If this is the idea, then I reject MI2, which says that perception is more fundamental than other forms of representation for appreciating the mind independence of objects. I agree that seeing a red tomato presents the subject with the tomato and its instances of redness, but I find it perfectly reasonable to deny that it presents them with the *mind independence* of the tomato. I understand what it is to experience tomatoes, their colors, and their shapes—I can attend to any of these entities in my experience—but I have no idea how to attend to the property of mind independence.

Conclusion The categorization account explains a wide range of data and escapes many important objections. But is it epistemologically adequate?

8.7 The Epistemology of Presentation

I am a philosophically savvy subject, one who possesses the concepts of presentation, mind independence, and mind dependence. When I see a blue ball in perfectly mundane conditions, I might form the belief that I am presented with a mind-independent ball. My belief amounts to knowledge. Now suppose that I suffer a matching hallucination and form the same belief: the belief that I am presented with a mind-independent ball. This belief does not amount to knowledge—it is not even true—but it is justified. Can the categorization theorist explain these facts? More generally, let a *presentational belief* be any belief of the form <I am presented with . . .>. Can the categorization theorist provide a satisfying epistemological account of presentational beliefs? That is the focal question of this section.

It is easy enough to see why my presentational belief would amount to knowledge when I am genuinely perceiving: the belief meets all of the standard conditions on knowledge. It is true, of course. It is also *safe*. Consider all nearby worlds in which I am presented with that same blue ball and form a belief on the same topic on the same basis—on the basis of whatever gestalt I find in my experience. In such worlds, I might see the ball from a slightly different vantage point. I will still automatically and subpersonally take my experience to have a perceptual gestalt; on that basis, I will categorize it as presenting a mind-independent ball, and either in virtue of that, or on that basis, I will form the belief that I am presented with a mind-independent ball. In every such world, my belief will be true.

My presentational belief will even be *sensitive*: in the nearest worlds in which I am *not* presented with that thing—say, worlds in which I turn my back on the ball or simply close my eyes—I will not believe that I am presented with a mind-independent ball. In addition, my presentational belief will be *formed via a reliable process*. For consider any process for forming beliefs on contingent, a posteriori matters. If the process is both safe and sensitive, it is hard to see how it could be unreliable.[21]

It may be objected that when I have a matching hallucination, I will form the same presentational belief. Might this undermine the status of my presentational belief as knowledge when I am genuinely perceiving? On pain of skepticism, it cannot. Perception surely positions me to know that I have hands, even though there are possible scenarios in which I form a similar but nonknowledgeable belief on the basis of a matching hallucination. There is no evident reason that my belief that I am presented with a mind-independent ball would be epistemically worse off than my belief that I have hands.

Now turn to the presentational belief that I form when I merely hallucinate a blue ball: again, the belief that I am presented with a mind-independent ball. This belief does not amount to knowledge: it is not even true. But it is plausible that the belief is justified. Can the categorization theorist explain why? One natural answer is that the hallucinatory presentational belief is the output of a reliable belief-forming process. Another natural answer is that a belief is justified as long as it is simply the endorsement of how things introspectively seem to the subject, in the absence of defeaters. The hallucinatory presentational belief would meet either condition.

Conclusion The categorization account explains a wide range of data, avoids a number of influential objections, and lets us tell a simple and attractive story about the epistemology of presentation. Still, it is too soon to accept it: there are two powerful rival accounts that we have not yet considered.

One rival account says that to have apparent presentational character is to have a causal content of a certain kind. Another rival account says that to have a particular kind of apparent presentational character is to have a content about a certain kind of subject independence and perspectival connection.

In the next two sections, I compare these rivals to the categorization account.

8.8 The Categorization Account versus the Causal Content Account

According to *existential representationalism*, all sensory experiences have a purely existential content concerning the objects and properties that are perceived.[22] For instance, when I sensorily imagine a yellow station wagon, my experience might have roughly the content <there exists a yellow station wagon in front of me now>. Throughout most of this book, I have been bracketing this theory because it violates chapter 1's framing assumption. However, I will briefly bring it back into the main discussion, for existential representationalists have developed a clever proposal—the *causal content account*—about the nature of apparent presentational character. The purpose of this section is to compare this account to the categorization account.

Here is the account in question:

> The **causal content account**: For a mental item to have apparent presentational character toward a target is for it to represent its target as standing in a certain causal relation to that very mental item.[23]

The thought is that when I *see* a yellow station wagon rather than imagining one, my experience has a different content: roughly, <there exists a yellow station wagon in front of me now that stands in such-and-such causal relation to this very experience>.

The causal content theory has an important virtue: it can explain all of the data gathered so far in this chapter. For instance, the causal content theorist can explain why certain experiences present their targets without introspectively seeming to do so: these experiences actually stand in the relevant causal relation to their targets, but their contents are silent about this. They can also explain why certain experiences introspectively seem to present their targets without actually doing so: these experiences have causal contents without actually standing in the relevant causal relations to anything.

However, that does not give the causal content theorist any advantage over the categorization theorist, who can explain all of these data as well. (The categorization account does introduce a novel posit, categorization, but there is overwhelming empirical reason to introduce that anyway.)

But I believe that the categorization account has two potential advantages, and one definite advantage, over the causal content account.

First, it seems challenging to provide a theory of representation that would let experiences have causal contents of the right kind: it seems hard to give a principled account of precisely how an experience might end up representing specific causal relations between its targets and itself. Second, it is often said that the causal content account is phenomenologically implausible. The idea is that the subject should have direct introspective access to any element of the content of their perceptions and hallucinations, at least when these are hard-conscious. But introspection does not seem to confront us with causal relations between objects and hard-conscious perceptions or hallucinations.

Causal content theorists are familiar with these objections and have offered various responses to them.[24] I will not retrace these dialectics. I will simply observe that these objections do not arise for the categorization account, insofar as it does not posit such sophisticated contents in the first place. These are the *potential* advantages of the categorization account.

What I wish to focus on is a *definite* advantage for the categorization account. To appreciate what this is, notice that when I see a yellow station wagon, my experience positions me to make *direct reference* to that object. (I am emphasizing *directness* of reference here, not *special knowledge* of reference.) Even more simply, I can *directly attend* to that particular object. The causal content theorist has trouble explaining how this is possible. She says that the content of my experience is purely existential with respect to the station wagon—it just concerns the *existence* of such a thing. How, then, can I make direct reference to *that particular wagon*? How can I so much as *attend* to it directly?

The categorization theorist can easily answer these questions. If they are a singular representationalist or a rich pluralist, they can say that perceptions have successful singular contents, contents with particulars just built into them. If they are a naïve realist, they can say that hard-conscious perceptions acquaint us with particulars. Either way, perceived particulars would be available for attention and direct reference.[25]

For these reasons, although I see some merit in the causal content account, I think that the categorization account is even better.

8.9 The Categorization Account versus the Perspectival Content Account

In this section, I evaluate another account, the **perspectival content account**, which is meant to explain certain phenomenological differences between typical perceptual experiences and typical sensations.

A core idea of the perspectival content account is that when I see a blue vase, the content of my experience is not just that there is a blue vase in front of me. It is also part of the content that if I change my perspective on the vase, the vase will not *thereby* move. (The vase could move *for other reasons*, but not merely in virtue of a change in perspective.) In addition, it is part of the content that this change in perspective would, all else equal, cause a change in the phenomenology of my experience: I would experience whatever is on the other side of the vase. These perspectival contents are meant to be properly *experiential*: they are found at the personal level rather than the subpersonal one and within the experience itself rather than in postexperiential processing.[26]

What primarily motivates this account is a comparison between, for example, a *perception* of a bright green light and a *sensation* of a bright green afterimage. Although these experiences are very similar, there will be an obvious phenomenological difference between them. The perspectival content account explains this. It lets us say that when I see a bright green light, I represent it as something that will not move simply because I move and as something that I can see from a different perspective, but when I have a mere sensation of a bright green afterimage, I do not represent the target of my experience in these ways.[27]

The full force of the motivation can be brought out with an exotic example:

> Suppose you are looking at a tiny doll. You take yourself to be in the usual sort of circumstance with respect to the doll. . . . Moreover, you are correct: you are seeing a doll. You even play with the doll a bit, putting it into the little hands of its owner and then back to a shelf in front of you. Then your attention moves on to other things.
>
> After an hour or so, however, something odd happens. You look back at the doll on the shelf and find that it seems to have lost its independence: it moves with movements of your head as if you were wearing a helmet with an imperceptible arm extending from the front, keeping the doll in your field of view. You hypothesize that someone has somehow attached the doll to your eyeglasses using a very thin string, without your knowing it.
>
> . . . now suppose that the strange sequence of visual experiences continues in an even stranger vein. You decide to test the eyeglass hypothesis by moving your eyes without moving your head, and you find that the doll seems to move with your eyes as well. It seems to be sensitive to the slightest eye movement. And things get even stranger. When you close your eyes, you continue having a visual experience as of a doll. And when you try, with your eyes open, to put an opaque

> object right in front of the doll to block it from your view, your visual experience persists in being a visual experience as of a doll. Overall, your experience of the doll comes to operate much like the experience of "seeing stars" from being hit on the head or from standing up too quickly. Just as nothing can occlude the "stars," nothing can occlude the "doll"; and just as you can "see stars" while you are seeing other things, so too you continue to see things in the normal way even when the "doll" won't leave your field of view. As with "seeing stars," the apparent position of the "doll" is highly sensitive to eye movement.[28]

The story begins with doll experiences that seem perception-like and ends with very similar experiences that seem sensation-like. What constitutes this difference? According to the perspectival content account, it is that at the beginning of the story, you experientially represent the doll as something that you could get a different perspective on; by the end of the story, you do not.

The account may be summed up like this:

The *perspectival content account*:

(i) Almost all typical visual perceptions represent the target as independent of the subject and as perspectivally connected to the experience. Almost all typical visual sensations do not do this.

(ii) This difference in content constitutes the obvious phenomenological difference between typical visual perceptions and typical visual sensations.[29]

Ultimately, I will reject both parts of this account. Still, for the sake of clarity, we should notice that (i) is, by itself, perfectly consistent with the categorization account: the categorization theorist can allow that there is this difference in content between typical visual experiences and typical visual sensations. We get a competitor to the categorization theorist only when we add claim (ii), which says that this difference in content *constitutes* the relevant phenomenological difference.

How, then, does the perspectival content account, taken as a whole, compare to the categorization account?

The categorization theorist can also explain what is happening in the exotic example of experiencing a doll. They will say that at the start of the story, you categorize your experience as presenting a mind-independent object. However, at the turning point of the story, the sensory qualities associated with your doll experience cease to interact with other sensory qualities in the usual ways. They do not change in the right ways as you move around, cover your eyes, or close your eyes. Instead, the sensory qualities associated with the doll persist no matter what—even while your experience has a perception-like gestalt with respect to other environmental objects. Thus, you begin to categorize your experience as presenting something mind dependent. Thus, the perspectival content account does not have any particular edge at handling the exotic example.

At the same time, the perspectival content account faces two important objections that the categorization account does not. The first objection will show us that the perspectival content account has gone wrong *somewhere*. The second objection will reveal the precise location of the mistake.

The first objection begins with the observation that, according to the perspectival content account, it is only *certain typical* visual experiences that have perspectival content. The account does *not* say that *all* visual experiences have such content. And rightly not: when we see the sky, we do not represent it as something that, if we changed our perspective on it, would cause a change in the phenomenology of our experiences of it.[30] However, this restriction in the scope of the account generates a problem. For compare an ordinary *perception* of a blue expanse of the sky, on the one hand, to a very similar, vivid *sensation* of a blue, sky-like expanse. There is an obvious difference between these two experiences. It is the same as the difference between seeming to see a green light and seeming to experience a green afterimage. But the perspectival content theorist cannot accept this: they say that the latter difference is constituted by a difference in perspectival contents, while the former difference is not. Indeed, they do not yet have *any* account of the difference between the sky perception and the sky-like sensation.

Something has gone wrong. What, though? My next objection will suggest an answer to this question. Grant for the sake of argument that experiences do have perspectival contents. Thus:

(+) When an ordinary subject sees a green light, they represent it as something that they can see from different perspectives and that does not move simply because they move.

(–) When an ordinary subject experiences a green afterimage, they do not represent it in these ways.

Now, our explanatory target is this:

(Difference) When an ordinary subject sees a green light, it introspectively seems *real* to them. When they experience a green afterimage, it introspectively seems *unreal* to them.

It is evident that (+) and (–) do not yet explain (Difference). For (Difference) is a claim about introspective appearances of *reality* and *unreality*, but (+) and (–) do not speak to these topics at all. To complete the explanation, we must add the following claim:

(Link) Representing a target as being visible from different perspectives, and as not moving simply because the subject moves, *constitutes* the introspective appearance that this target is real. Representing a target as not having these perspectival features *constitutes* the introspective appearance that this target is unreal.

My objection is that (Link) is false. Representing a target as having (or lacking) various perspectival features may well *cause* an ordinary subject to take the target as real (or unreal). But it does not *constitute* it.

To appreciate this, imagine a genetically modified human being who can see their own retina. Perhaps their retina changes color periodically, and these changes in color are clearly visible to them—unlike us, they have no trouble focusing on something so close up. (Or just imagine that they are wearing a nonslip contact lens with a visible speck of dirt on it.) Of course, no matter how much this subject moves their eyes, they cannot change their perspective on their retina itself. In addition, their retina moves simply because they move their eyes. So their retina lacks the relevant perspectival features. Suppose now that our subject correctly experientially represents their retina in these ways—they represent it *as* lacking the perspectival features. Does this *make it the case* that they represent their retina as unreal? Obviously not. As they see their retina change colors from deep blue to light green, their experience might perfectly well seem perception-like rather than sensation-like.

At this point, the perspectival content theorist might beat a retreat. They might grant that representing the perspectival features does not *constitute* the introspective appearance of reality that is distinctive of perception. But, they might say, it *causes* that introspective appearance. Fine! But then there is a hole in their account: we still need to know what *does* constitute the introspective appearance of reality. I propose to fill the hole with the categorization account, which says that we just directly categorize targets as mind independent or mind dependent (or, if you like, as real or unreal).

This has a further effect. Once we have the categorization account to explain what constitutes the introspective appearance of reality, we no longer have a good reason to think that the relevant perspectival features are represented *at the personal, experiential level.* The perspectival content theorist's earlier motivation for accepting this claim might be put like this ("PER" stands for "personal-level, experiential representation"):

PER1. The sense of reality is found at the personal, experiential level.

PER2. The sense of reality is constituted by representations of the perspectival features.

PER3. Representations of the perspectival features are found at the personal, experiential level.

But we have now discarded PER2. We thereby lose the case for PER3. To come at the point from the other end, we would still get a sound explanation of the phenomenological difference between perception and sensation merely by saying this (where "SNR" stands for "subpersonal, nonexperiential representation"):

SNR1. There are *subpersonal, nonexperiential* representations of the perspectival features.

SNR2. These often *cause* categorizations of targets as mind independent (real) or mind dependent (unreal).[31]

Thus, we even lose the support for part (i) of the perspectival content account—the part that says that typical visual perceptions represent perspectival features *at the personal level,* and typical visual sensations do not.

Let us step back. There are deep similarities between the perspectival content account and the categorization account. Both are motivated by the explanatory need to distinguish perceptual experiences from mere sensations and by the observation that the subject is typically sensitive, in some sense, to certain sophisticated differences between the two. So far I am on board. Where I think the perspectival content account goes wrong is in trying to pack these sophisticated differences *into the very content of experience.* The categorization theorist instead tucks the sophisticated differences into the *basis* that our subpersonal systems use for representing our experiences in certain ways, while keeping the actual contents of the representations very simple.

Conclusion At this point, the categorization account has survived a thorough examination: it explains a wide range of data, avoids many important objections, and compares favorably to all of its main rivals. We can safely accept it.

What consequences does this have?

8.10 The Significance of the Categorization Account

In this section, I make a series of observations about how the categorization account bears on larger questions about theories of perception.

First observation Which theories can profitably be combined with the categorization account? I have already shown that the full explanatory benefits of the account are available to rich pluralists and representationalists. The more interesting question is about naïve realists.

Naïve realists can make straightforward use of the categorization account to explain why hard-conscious *perceptions* have apparent presentational character. But what about hard-conscious *hallucinations*—why do these typically have apparent presentational character? The categorization account says that it is because they typically have a perceptual gestalt. Whether the naïve realist can take this on board depends on which theory of hallucination they accept.

In chapter 6, I ruled out all but three naïve realist theories of hallucination. One surviving theory was flexible naïve realism, which says that hallucinations involve acquaintance with mind-dependent instances of the sensory qualities. A second survivor was the quality-awareness view, which says that what it is to hallucinate is just to be acquainted with the sensory qualities, understood as universals, *without* being acquainted with any particulars. Both of these theories can allow that hallucinations typically have perceptual gestalts: these theories say that, like hard-conscious perceptions, hard-conscious hallucinations involve genuine awareness of either instances of the sensory qualities or sensory qualities themselves. Thus, flexible naïve realism and the quality-awareness view can be happily conjoined with the categorization account.

The third surviving theory of hallucination was the imaginative view, which says that part of what it is to be a hallucination is to be an involuntary sensory imagining. Because there are many different theories of sensory imagination, it is hard to assess whether this view can explain why hallucinations can have the perceptual gestalts. I will simply flag this as an open question.

In any case, my first observation is that rich pluralists, representationalists, and naïve realists who accept certain theories of hallucination can all enjoy the full benefits of the categorization account.

Second observation Does the categorization account favor any theory in particular? In one sense, my answer is *no*; in another sense, *yes*.

On the one hand, the categorization account, with all of its explanatory benefits, is available to rich pluralists, representationalists, and naïve realists. In this sense, the account gives equal favor to all of these theories.

On the other hand, the fourth major motivation for naïve realism is the thought that this theory has the best account of the apparent presentational character of perception (though not of hallucination). Supposedly, other theories cannot do justice to this phenomenological feature. Thus, if the best account of apparent presentational character is in fact available to all, then this supposedly central advantage of naïve realism evaporates, and the scales tip toward rival theories. In this sense, I think that the categorization account favors rich pluralism and representationalism over naïve realism.

Third observation In chapter 4, while considering an ordinary case in which you see a scarlet pepper, I mentioned these questions:

Q10. Why does your perception of the scarlet pepper have the phenomenology associated with perception rather than having the phenomenology associated with sensory imagination?

Q11. Why does your perception of the scarlet pepper seem to be an experience of a mind-independent particular rather than merely an experience of a mind-dependent particular or an experience of universals?

I promised that I would later argue that these questions do not constitute *novel* hard problems of consciousness—problems over and above the ones constituted by questions like these:

Q3. Why do you experience scarlet rather than not experiencing anything?

Q4. Why do you experience scarlet rather than experiencing lime green?

I will now make good on that promise.

This chapter has defended a fully reductive analysis of apparent presentational character. When I apply this analysis to my apparently presentational experiences, I cannot find any lingering hard problem about why they appear to *present* their targets: I find the analysis itself fully satisfactory, and I can understand exactly how it is possible for a purely physical being to satisfy the conditions of the analysis. To be sure, there are hard problems surrounding why my apparently presentational experiences make me aware of the sensory qualities that they do. There might also be hard problems surrounding intentionality and thus surrounding the existence of concepts. But neither of these hard problems has anything to do with apparent presentational character per se. Modulo these hard problems, I think I understand exactly how it is possible for a purely physical being to enjoy apparent presentational character.[32]

Summary I have made three observations. First, that the categorization account, with all of its explanatory benefits, is equally available to rich pluralists, representationalists, and certain kinds of naïve realists. Second and consequently, that all of these theorists have equal access to the best account of apparent presentational character—but this removes a major supposed advantage of naïve realism. Third, that the categorization account supports the idea—which I drew on in chapter 4—that apparent presentational character does not, in and of itself, generate any novel hard problems of consciousness.

8.11 Conclusion

In this chapter, I have argued that the best account of apparent presentational character is the categorization account. This account tells us both what *constitutes* apparent presentational character and what *causes* it in typical, mature human subjects. The constitutive portion of the account is very simple: apparent presentational character is just a matter of categorizing—applying the relevant concepts to—the experience and its targets. It is perfectly transparent why this would constitute apparent presentational character. As for what *causes* apparent presentational character, most accounts try to find *individual* features that are necessary or sufficient for this, but this turns out to be very hard to do. A great advantage of the categorization account is that it instead appeals to an *overall gestalt* of features. This explains the complex pattern of cases in which apparent presentational character does and does not occur.

That, at last, removes the fourth and final motivation for naïve realism. Naïve realists do *not* have unique access to the best account of apparent presentational character. The best account—the categorization account—is equally available to naïve realists, rich pluralists, and representationalists.

Of the four original motivations for naïve realism, none remain.

Still, it is not yet clear how naïve realism ultimately compares to rich pluralism. In recent chapters, I have argued that both theories, at least in their best forms, are compatible with our best accounts of intrinsic awareness, strong referential character, presentation, and apparent presentational character. And in earlier chapters, I argued for extended partial revelation, but I showed that this thesis, too, is compatible with both theories. More investigation is required to break the tie.

Thus, the task of the next chapter is to make a final evaluation of naïve realism.

9 A Final Assessment of Naïve Realism

We can now make our final comparison between naïve realism and rich pluralism.

Pulling together our earlier conclusions, I will suggest that naïve realism has no significant advantages over rich pluralism (section 9.1). However, I will argue that rich pluralism has significant advantages over all of the naïve realist theories that remain on the table. It is better than the remaining monistic forms of naïve realism, namely flexible naïve realism (section 9.2), the imaginative view (section 9.3), and the quality-awareness view (section 9.4). It is also superior to pluralistic forms of naïve realism (section 9.5) and, for that matter, to one last, strange pluralistic theory (section 9.6).

For these reasons, we should prefer rich pluralism to any form of naïve realism (and to the strange pluralistic theory). And, as I argued in part II, we should prefer rich pluralism to any form of representationalism. I conclude, with some firmness, that rich pluralism is the best theory of perception to date (section 9.7).

9.1 Rich Pluralism versus Naïve Realism: The Current State of Play

Let us remind ourselves of the current dialectic between rich pluralism and naïve realism.

In chapters 2–4, I defended extended partial revelation, thereby removing many naïve realist theories from contention. In addition, in section 6.2 of chapter 6, I argued that most versions of naïve realism cannot explain why hard-conscious hallucinations partly reveal the intrinsic characters, and the essences, of the sensory qualities. However, a handful of naïve realist theories survived these challenges. These included three *monistic* naïve realist theories—flexible naïve realism, the imaginative theory, and the quality-awareness theory—as well as *pluralistic* naïve realism. These are the versions of naïve realism that I will examine in this chapter.

In chapters 6–8, I also considered whether these naïve realist theories have any advantages over rich pluralism.

Naïve realists have rested their theory on four core motivations. They claim to have the best theory of, first, why hard-conscious perception reveals the intrinsic characters of things; second, why it has strong referential character; and, third and fourth, why it presents its targets, and why it (correctly) introspectively seems to do so. But in chapters 6–8, all four of these motivations were swept away. I argued that naïve realists have missed much of the data, and once these are laid out in full, the best accounts turn out to be available to theorists of all kinds: not just naïve realists but also rich pluralists and representationalists.

Nor is naïve realism more parsimonious than rich pluralism. Pluralistic naïve realism obviously makes no gains in parsimony. As for monistic naïve realists, I argued that these theorists should say that acquaintance is partly revelatory and that revelatory awareness is what generates the hard problems of consciousness and the external world (chapters 2–4). But once the monistic naïve realist accepts these commitments, they must concede that acquaintance is by its nature hard-conscious—that is the only way to correctly predict which mental events generate hard problems. So their main posit, acquaintance, cannot explain the nature of hard-*un*conscious perceptions; they must posit a second kind of awareness, *covert awareness*, to explain these. That makes their theory just as complex as rich pluralism, which also relies on two core posits: deep awareness and sensory representation (section 5.6 in chapter 5).

So far, then, we have found no particular advantages for naïve realism over rich pluralism (or, for that matter, for naïve realism over representationalism). But neither have we found any particular advantages running in the opposite direction. Thus, across the next four sections, I propose to make a closer examination of each of the naïve realist theories that remain: flexible naïve realism (section 9.2), the imaginative theory (section 9.3), the quality-awareness theory (section 9.4), and pluralistic naïve realism (section 9.5).

9.2 Problems for Flexible Naïve Realism

Like any naïve realist, the *flexible naïve realist* holds that to perceive a red tomato is to stand in a primitive relation of awareness, *acquaintance*, to one or more instances of redness in a mind-independent tomato. However, the flexible naïve realist also thinks that to *hallucinate* a red tomato is to be acquainted with one or more instances of redness. How is this possible?

Their idea is that sensory qualities are *ontologically flexible*: the existence of an instance of redness can be fully grounded *either* in its inhering in an ordinary object such as a tomato, *or* in a subject's being in a certain internal state (perhaps a brain state). Thus, the flexible naïve realist says that when I hallucinate a red tomato, I am aware of a perfectly real but mind-*dependent* instance of redness. This instance of redness exists purely in virtue of my being in a certain internal state, and the instance does not inhere in anything—not even my mind or experience. But what happens when I actually see a red tomato? Then I am acquainted with an instance of redness whose existence is *ontologically overdetermined*: its existence is fully grounded in the fact that I am acquainted with it *and* in the fact that it inheres in the tomato.[1]

Flexible naïve realism has many virtues. It accommodates the attractive idea that perceptions of redness and hallucinations of redness are deeply similar, for it says that both involve genuine awareness of instances of redness. It honors the powerful intuition that we are genuinely presented with *instances* of redness, even when we are hallucinating, for it analyzes presentation in terms of acquaintance. (Even rich pluralists cannot say this. That is one point for flexible naïve realism.) At the same time, it retains the natural idea that tomatoes can be red, since it says that redness is ontologically flexible and thus capable of inhering in tomatoes. It even accommodates the idea that the tomato that I see is *mind-independently* red. For that very instance of redness could, and likely will, continue to exist whether or not there is anyone to perceive it. Its inherence in the tomato is enough to guarantee its existence.

I believe that there is something brilliant in the way that flexible naïve realism is constructed. I prefer it to typical versions of naïve realism. Still, I think the theory faces a problem of substance and a problem of methodology.

The problem of substance is this. When I hallucinate a red tomato, the flexible naïve realist says that I am acquainted with a real, albeit mind-dependent, instance of redness. This instance of redness really exists, they add, but it does not inhere in anything whatsoever. I cannot make sense of this. I can understand how a *universal* might be uninstantiated, but how can an *instance* of redness not inhere in anything—not an object or even a region of space?

This brings us to a second problem, one of methodology. Flexible naïve realism includes a number of elaborate commitments. I am perfectly willing to accept such commitments *if there is specific evidence that favors them over rival commitments*. But flexible naïve realism is deliberately constructed so as to remove the possibility of such evidence.

For instance, you might have thought to test flexible naïve realism by considering what would happen in the following case. A subject sees a tomato; the tomato is destroyed; and at that very moment, the subject's brain is artificially stimulated to produce an experience of redness. Flexible naïve realism would seem to predict that a single instance of redness would continue to exist throughout this story. For if the existence of the instance of redness is ontologically overdetermined, then even when *one* of its ontological supports, the tomato, is taken away, the instance can persist due to the presence of its *other* ontological support, the subject's internal state. Moreover, since the instance of redness was publicly visible to begin with, it should continue to be publicly visible if it continues to exist. *That* would certainly be a distinctive (albeit false) prediction.

In fact, however, the flexible naïve realist avoids making this prediction. They say that the subject enters a new internal state as soon as the tomato is destroyed: the subject's internal state when they see a tomato is individuated *by that very tomato*. Thus, once the tomato is destroyed, the subject enters a qualitatively identical but numerically distinct internal state. The result is that even though the instance of redness that inheres in the tomato is ontologically flexible—even though the existence of *that very instance* is fully grounded in the subject's internal state—it is *metaphysically impossible* for that instance to exist in the absence of the tomato, purely in virtue of the subject's being in that internal state.[2] (This is, by the way, yet another elaborate commitment of the view.)

This is a conspiracy theory: whenever the theory *would* deliver distinctive predictions, further commitments are added to cancel those predictions. It is possible that there are metaphysical conspiracies of this sort. But we would need a lot of evidence before we could rationally believe in them. Rich pluralism, by contrast, makes a number of bold and specific predictions, which, lo and behold, turn out to be correct.

Conclusion Flexible naïve realism has one important advantage over rich pluralism: it allows that even in hallucinations, we are sensorily aware of *instances* of sensory qualities. However, the view comes with large costs. First, it posits instances of redness that do not inhere in anything, not even a region of space. Second, it posits metaphysical conspiracies: it adds new commitments to cancel any distinctive predictions that it would otherwise make. Thus, all things considered, I believe that rich pluralism is the better theory.

But let us consider another naïve realist option.

9.3 Problems for the Imaginative View

Some naïve realists have proposed the *imaginative view*, which says that part of what it is to be a hallucination is to be an involuntary sensory imagining. This is not intended as a full account of the nature of hallucination: it is granted that some involuntary sensory imaginings are not hallucinations.[3]

I see several weaknesses in this account.

First, as we saw in chapter 6, hard-conscious perceptions reveal to us the intrinsic characters of the sensory qualities. I argued that, to explain this, the naïve realist should say that hard-conscious perceptions acquaint us with particulars and their sensory qualities (or instances thereof), and the relation of acquaintance always reveals the intrinsic characters of its targets. However, I also observed that hard-conscious *hallucinations* reveal to us the intrinsic characters of the sensory qualities. How will the imaginative theorist explain this?

Given their commitments, they should say that *sensory imagination* in general reveals the intrinsic characters of the sensory qualities. However, this just pushes back the problem. It remains to be explained how sensory imagination can do this. There are two main proposals in the literature.

The first proposal is that sensory imagination acquaints us with sensory qualities but not with any particulars. This is simply a version of the quality-awareness view, which I will discuss in the next section.

The second proposal is that sensory imaginings are representations *of perceptions*.[4] The imaginative theorist already has an account of how hard-conscious perceptions of round tomatoes reveal to us the intrinsic character of roundness. They might add that in virtue of being representations of perceptions, hard-conscious sensory imaginings have the same revelatory power.

However, this view comes with an important cost. To see what it is, notice that a sensory imagining cannot be just *any* kind of representation of a perception. We can entertain abstract thoughts about perceptions, and thereby represent perceptions, without sensorily imagining anything. It seems that sensory imaginings must be understood as representations of perceptions that are distinctively *sensory*. Thus, just like the rich pluralist, the imaginative theorist posits, first of all, *sensory representations* of some kind to account for hallucinations. But, second, the imaginative theorist posits *acquaintance* to account for perceptions and, third, *covert awareness* to account for hard-unconscious perceptions. This makes their theory less parsimonious than rich pluralism, which posits only deep awareness and sensory representation.

But wait—does the imaginative theory have an advantage in explanatory power, insofar as it also includes an account of sensory imagination? No. The rich pluralist can account for sensory imagination without adding any new posits. They can say that sensory imagination is a particular kind of sensory representation and that it is hard-conscious just in case it is deployed in concert with deep awareness.

I see a further problem for the imaginative theorist. The subjects of hallucination often take themselves to be perceiving. Thus, if the imaginative view is true—if hallucinations are a type of sensory imagination—then these subjects confuse *imaginative* experiences with perceptions. This is surprising: we do not usually have trouble distinguishing imagination from perception. What is different in these cases?

The imaginative theorist might reply that many hallucinations occur in circumstances that impair the subject's rationality. Hallucinations are often produced by drugs, and perhaps they also occur when we dream. In these cases, the subject might mistake a sensory imagining for a perception simply because they are being irrational. But plenty of hallucinations, such as those involved in Charles Bonnet syndrome, occur in subjects who do not seem to be suffering from any rational impairment. In these cases, there is strong pressure to grant that the subject's hallucinatory (and allegedly imaginative) experience is *in principle* indiscriminable from a matching veridical perception. It is not clear why this would be the case, given that the hard characters of these experiences are supposed to be radically different: supposedly, the hard character of the perception is constituted by worldly objects and property-instances, the hard character of the hallucination by something entirely different.

There is one last problem. It is usually granted that at least in a normal subject, holding fixed their history, undergoing the right kind of purely neural event is sufficient to produce hallucinations. And the imaginative theorist identifies hallucinations with sensory imaginings. Thus, they should say that for a normal subject, undergoing the right kind of purely neural event is sufficient to produce sensory imaginings, as well. But whatever kind of purely neural event occurs in hallucination, it will occur in a matching perception, too. Thus, the matching perception ends up having too many hard characters—*both* the one associated with the purely neural event *and* the one constituted by the subject's being nonrepresentationally aware of objects and property-instances in the world. That is an implausible result.

The imaginative theorist can avoid this result by suggesting that for a sensory imagining to occur, the subject must *fail* to be acquainted with whatever they experience.[5] But look at the resulting view. It says that when you see a scarlet pepper, your experience has a certain hard character, call it H_1, because you undergo a certain neural event that helps acquaints you with that very pepper. H_1 is constituted by worldly objects and property-instances. The view adds that when you undergo that same neural event without being acquainted with anything, you sensorily imagine a pepper. Your experience has a hard character, H_2, that is not constituted by any worldly objects or property-instances. Thus, H_1 and H_2 are radically different—but the latter is in principle indiscriminable from the former.

This analysis is needlessly complex. A much simpler view is that given the subject's history, the occurrence of the neural event in them—whether that event constitutes a perception, a hallucination, or a sensory imagining—suffices, at least nomologically, for the occurrence of an experience with a single hard character. This simpler view is available to the rich pluralist.

Conclusion Naïve realism, if it is conjoined with the imaginative view, has several weaknesses when measured against rich pluralism. First, the imaginative naïve realist ends up positing three very different kinds of mental events—acquaintance, covert awareness, and sensory representation—to account for perceptions, hallucinations, and sensory imaginings. The rich pluralist can do the work with only two posits: deep awareness and sensory representation. Second, the imaginative naïve realist says that it can be impossible in principle to discriminate the hard characters of perceptions and hallucinations, *even though* these hard characters are of radically different kinds—the former is constituted by worldly objects and property-instances, the latter by something entirely different. The rich pluralist can just say that these hard characters are indiscriminable because they are the same. Finally, the imaginative theorist must say these two in-principle indiscriminable hard characters just so happen to be underpinned by a common neural event, *despite* their radical differences in kind. The rich pluralist can accept the simpler view that the common neural event underpins a single kind of hard character.

In the final analysis, then, the imaginative view has many costs, but no benefits, when measured against rich pluralism.

9.4 Problems for the Quality-Awareness View

The *quality-awareness view* says that what it is to perceive is to be acquainted with particulars instantiating sensory qualities (understood as universals), and that what it is to hallucinate is to be acquainted with sensory qualities alone, *without* being acquainted with any particulars.[6] In this section, I will argue that although the quality-awareness view shares the explanatory power of rich pluralism, it comes with a surprising additional cost.

The explanatory power of the quality-awareness view The quality-awareness theorist has access to many of the explanatory benefits of rich pluralism. To begin with, they can say that acquaintance always reveals the intrinsic characters of its targets—even hard-conscious hallucinations, they can add, reveal the intrinsic characters of the sensory qualities. These claims are plausible in and of themselves (see chapter 6); they potentially let us explain why experience reveals the very essences of the sensory qualities, without revealing the essences of objects or events (see appendix 2A of chapter 2); and they are part of the best accounts of strong referential character and presentation, which the quality-awareness theorist is welcome to accept (see chapter 7).

The quality-awareness view has one last benefit. As I argued in chapter 5, experiences never err about the essences of the sensory qualities. If we say that experiences *represent* the sensory qualities, it becomes very hard to explain this: representations are, by definition, subject to error. The quality-awareness theorist does better. They say that experiences *acquaint* us with the sensory qualities, where acquaintance is a relation that is immune to error. This delivers the desired result that experience never errs about the essences of the sensory qualities.

In sum, the quality-awareness theorist can hijack many of the core motivations for rich pluralism. However, I will argue that the theory comes with an important and unexpected cost.

A cost for the quality-awareness view Surprisingly, I believe that any adequate version of the quality-awareness theory will be *more* complex than rich pluralism. It will take some time to develop this point.

We can start by asking, How will the quality-awareness theorist account for hard-*un*conscious perceptions? They are a monist, and as I argued in chapters 2–4, the only viable form of monism is impure revelatory monism. I went on to argue in chapter 5 that any impure revelatory monist should hold that sensing (which, for the quality-awareness theorist, is simply acquaintance) is, by its nature, hard-conscious; thus, it is of no use in accounting for hard-*un*conscious perceptions. The quality-awareness theorist must therefore say that hard-unconscious perceptions involve a novel kind of awareness, call it *covert awareness*. As a result, just like the rich pluralist, the quality-awareness theorist posits at least two very different kinds of awareness.

But things get worse. The quality-awareness theorist must also account for hard-conscious and hard-unconscious *hallucinations*. In both cases, *at some level*, the subject takes the particulars that they hallucinate to exist. This is obvious for hard-conscious hallucinations: when I hallucinate a scarlet pepper, at some level, it seems to me that "that" particular pepper exists—even if I happen to know better in the end. But this also occurs in hard-unconscious hallucinations. Consider creatures who lack hard-consciousness entirely: even they can hallucinate particular peppers, in a way that will show up in their thoughts and behaviors. Something similar would occur in hard-unconscious hallucinations within our dorsal systems, although here it is perhaps only the dorsal system, rather than the subject, that takes the hallucinated particulars to exist.

Thus, in both hard-conscious *and* hard-unconscious hallucinations, the hallucinated particulars are, at some level, taken to exist. I ask the quality-awareness theorist: in virtue of what does this occur?

They can readily answer this question for *perceptions*. If the perception is *hard-conscious*, then the particulars are taken to exist because we are *acquainted* with them; if the perception is *hard-unconscious*, the particulars are taken to exist because we are *covertly aware* of them. But these explanations cannot be extended to hallucinations. The hallucinating subject is neither acquainted with, nor covertly aware of, any particulars.

Some quality-awareness theorists have said that a hallucinating subject takes the hallucinated particulars to exist because they are aware of sensory qualities that include *changing spatiotemporal properties*. For instance, suppose that a subject hallucinates a scarlet pepper rolling toward them. Then they are genuinely aware, whether via acquaintance or covert awareness, of a series of qualities: *being scarlet and pepper-shaped and 1 meter away from me right now, being scarlet and pepper-shaped and 0.9 meters away from me right now, being scarlet and pepper-shaped and 0.8 meters away from me right now*, and so on.[7] This fools them into thinking that there exists an actual scarlet pepper.

This account, even if it is true, does not answer our question. At best, the account tells us what *causes* us to take hallucinated particulars to exist. But we want to know what *constitutes* this.

Another proposal is that for a hallucinating subject to take the hallucinated particulars to exist is simply for them to *believe*, on the basis of their hallucination, that those particulars exist. But this cannot be right. One problem is that the subject might know that they are hallucinating and thus not believe that the relevant particulars exist, but their hallucination will still seem to make them sensorily aware of particulars. Another problem, and to my mind a deeper one, is that taking particulars to exist in hallucination is plausibly a sensory matter rather than a postsensory, cognitive one.

We are running out of options. As far as I can tell, the only reasonable resort for the quality-awareness theorist is to posit an entirely new kind of mental event, call it *erring*, that *constitutes* taking particulars to exist in cases of hallucination. The idea would be that erring is a mental event that is not only mistaken with respect to particulars but that has no successful counterparts. In that respect, erring would contrast with representations, which, by their very natures, can come in successful and unsuccessful varieties.

The bottom line is that to account for the full range of perceptions and hallucinations, the quality-awareness theorist should posit *three* entirely separate kinds of mental events: *acquaintance*, which constitutes hard-conscious perceptions (and our relation to the sensory qualities in hard-conscious hallucinations); *covert awareness*, which constitutes our awareness in hard-unconscious perceptions; and *erring*, which constitutes our taking particulars to exist in hard-conscious and hard-unconscious hallucinations.

Conclusion The quality-awareness theory has great explanatory power, but the price of admission is three basic kinds of mental events. Rich pluralism lifts the same explanatory load with just two basic kinds of mental events: deep awareness and sensory representation. That makes it the better theory.

Our examination of *monistic* versions of naïve realism is now complete: rich pluralism is preferable to any of them. Let us next consider *pluralistic* naïve realism.

9.5 Problems for Pluralistic Naïve Realism

In this section, I argue that rich pluralism is superior to pluralistic naïve realism. I will be brief, since we have already encountered the key ideas.

Like rich pluralism, pluralistic naïve realism says that hard consciousness is constituted by *deep awareness*, understood as a partly revelatory form of awareness whose only targets are the sensory qualities. However, unlike rich pluralism, pluralistic naïve realism accounts for perceptions, including hard-unconscious ones, in terms of a nonrevelatory *and nonrepresentational* relation of sensory awareness—call it *acquaintance*—to particulars. The rich pluralist thinks that this form of awareness is *representational* instead.

Pluralistic naïve realism has many advantages. First, it can account for hard-unconscious perceptions (a central topic of chapter 5) as cases of acquaintance without deep awareness. In addition, pluralistic naïve realism accepts the very same theory of deep awareness as rich pluralism, which gives it great explanatory power. This theory can explain why extended partial revelation is true: it is because the partly revelatory relation of deep awareness only ever targets the sensory qualities (see chapters 2–4). Extended partial revelation can in turn explain why experience partly reveals the intrinsic characters of the sensory qualities (see chapter 6), a claim that itself provides access to the best accounts of presentation and strong referential character (see chapter 7). Additionally, the pluralistic naïve realist's theory of deep awareness can explain why experience never errs about the essences of the sensory qualities: it is because deep awareness is a nonrepresentational relation and is thus immune from error (see section 5.5 in chapter 5).

However, pluralistic naïve realism shares a major weakness with the quality-awareness theory. In hallucinations, we take various particulars to exist. In virtue of what does that occur? For precisely the reasons discussed in the previous section, this cannot be explained in terms of (i) acquaintance or covert awareness, (ii) awareness of changing spatiotemporal properties, or (iii) beliefs about particulars. The pluralistic naïve realist's best hope is to posit a new kind of mental event, *erring*, that not only is mistaken with respect to particulars but also has no successful counterparts.

A final problem arises for sensory imaginings. (It is reminiscent of a problem that came up for the imaginative view.) It is surely metaphysically possible for sensory imaginings to occur without hard consciousness. What *are* hard-unconscious sensory imaginings? They are not cases of acquaintance with particulars, since they are imaginings. Nor are they cases of deep awareness, since they are hard-unconscious. The natural move, then, is for the pluralistic naïve realist to say that hard-unconscious sensory imaginings are *representations*, of a distinctively sensory kind, of perceptions.[8] Hard-*conscious* sensory imaginings can then be understood as sensory representations of this kind that occur in concert with deep awareness. The lesson is that the pluralistic naïve realist will need to posit something new—perhaps a novel kind of sensory representation—*in addition to* the relations of acquaintance and deep awareness.

Conclusion Like rich pluralists, pluralistic naïve realists can shoulder the full explanatory load. The problem is that, to do this, they must posit no fewer than four different kinds of mental events: not just (i) acquaintance and (ii) deep awareness, but also (iii) erring, to explain why hallucinations seem to make us aware of particulars, and (iv) sensory representations, to account for hard-unconscious imaginings. Rich pluralism does the same work while positing only sensory representation and deep awareness. It is the better view.

If rich pluralism is superior to all monistic versions of naïve realism, as I argued in sections 9.2–9.4, and to pluralistic naïve realism, as we have just concluded, then it is better than any naïve realist theory of perception.

Still, there is one last view to consider.

9.6 Problems for Strange Pluralism

Thus far, I have argued that rich pluralism is superior to pluralistic representationalism and pluralistic naïve realism. But there remains one final pluralistic rival, albeit a very odd one, to rich pluralism. The purpose of this section is to examine it. I will be extremely brief.

The view in question, call it *strange pluralism*, says that to perceive is to deploy two very different kinds of awareness: partly revelatory *but representational* awareness of sensory qualities, together with nonrevelatory *and nonrepresentational* acquaintance with particulars. The italicized phrases in the last sentence mark the key points of departure from rich pluralism.

Strange pluralism inherits all of the defects of pluralistic representationalism and pluralistic naïve realism. On the one side, like pluralistic representationalism, strange pluralism predicts that it is possible to err about the very essences of the sensory qualities. I argued in chapter 5 that this is false. On the other side, like pluralistic naïve realism, the strange pluralist must posit some novel kind of mental event, *erring*, to explain why hallucinating subjects take hallucinated particulars to exist. And, like the imaginative theorist, they must posit another novel kind of mental event, perhaps some form of *sensory representation*, to account for hard-unconscious sensory imaginings. This brings their total number of posits to four.

Yet I can find no merits of this view in comparison with other forms of pluralism. Thus, strange pluralism can be swiftly dismissed.

9.7 Conclusion

This concludes my case for rich pluralism against any rival theory of perception. This case has had many parts. Let us snap them into place.

Rich pluralism versus representationalism I began in chapters 2–4 by identifying the best versions of representationalism: impure revelatory representationalism (the best monistic version) and pluralistic representationalism (the best pluralistic version).

We found that these theories had no notable advantages over rich pluralism. The rich pluralist could easily honor all of the motivations for representationalism, since they accepted all of the core representationalist posits—they just added a new posit about deep awareness. And these representationalist theories ended up being just as complex as rich pluralism. This was obvious for pluralistic representationalism, but it was also true of impure revelatory representationalism, which needed to posit a new kind of mental event, *covert awareness*, to account for hard-unconscious perceptions and hallucinations.

Yet we saw in chapter 5 that rich pluralism had advantages that were denied to its representationalist rivals. Most importantly, both of these representationalist theories delivered the incorrect prediction that we can err about the very essences of the sensory qualities. In addition, impure revelatory monists were unable to identify any unifying factor across hard-conscious and hard-unconscious perceptions.

On these grounds, I concluded that rich pluralism was better than any version of representationalism.

Rich pluralism versus naïve realism In chapters 2–4, I also identified the best versions of naïve realism: impure revelatory naïve realism (the best monistic version) and pluralistic representationalism (the best pluralistic version).

I went on to undermine the four core motivations for naïve realist theories. I granted that, yes, hard-conscious perceptions reveal to us the intrinsic characters of things, but I showed that representationalists and rich pluralists can incorporate this claim into their theories just as easily as naïve realists (chapter 6). That gave all of these theories equal access to the best accounts of strong referential character and presentation (chapter 7). Meanwhile, the best account of apparent presentational character, the categorization account, was available to all (chapter 8). I concluded that naïve realists enjoy no comparative advantages.

But they do suffer from comparative disadvantages. Most provide inadequate theories of hallucination (chapter 6). Only three *monistic* versions of naïve realism survived this criticism. One was flexible naïve realism, but this theory posits instances of redness that do not inhere in anything, not even a region of space; it is extremely complicated; and it posits metaphysical conspiracies (section 9.2). A second option was the imaginative theory. However, this theory must posit three very different kinds of mental events; it says that it can be impossible in principle to discriminate hard characters that are of radically different kinds, and it adds that these in-principle indiscriminable hard characters just so happen to be underpinned by a common neural event (section 9.3). The final monistic version of naïve realism was the quality-awareness theory, which has to rely on three basic posits to the rich pluralist's two (section 9.4).

So much for monistic versions of naïve realism. Then there was pluralistic naïve realism, which has to posit *four* different kinds of mental events to account for the full range of perceptions, hallucinations, and sensory imaginings (section 9.5).

From this, I concluded that rich pluralism is better than any version of naïve realism.

Rich pluralism versus strange pluralism For the sake of completeness, I considered one last pluralistic view, strange pluralism. This view has nothing to recommend it: it suffers from all of the flaws of pluralistic representationalism and pluralistic naïve realism, while offering nothing in return.

Conclusion At this point, all promising rivals have been eliminated. We can conclude with some confidence that rich pluralism is the best theory of perception. That concludes this part of the book, part III.

But our work is not yet done. There remain some further explanatory challenges—challenges that are hard to meet on *any* theory of perception. I discuss these in the final part of the book, part IV.

IV Other Explanatory Challenges; Concluding Thoughts

In parts I–III, I have, for the most part, been playing offense: I have been identifying respects in which rich pluralism is better than its rivals. In this final part of the book, I switch to defense.

In particular, I confront two remaining explanatory challenges—challenges that are difficult to meet for any theory of perception. The first challenge, which I tackle in chapter 10, is to answer a deep dilemma about the phenomenal character of perception. The second challenge, which I meet in chapter 11, is to respond to a series of arguments from hallucination.

Some of my answers to these challenges will also be available to my representationalist and naïve realist rivals. That is fine: now that I am in my defensive stance, I am not trying to find more comparative advantages for rich pluralism. I am just protecting the theory from attack.

Once that has been done, chapter 12 will cap things off: I will sum up the entire book's arguments for rich pluralism. Then I will rest my case, and let you be the judge.

10 Phenomenal Character

So far, the rich pluralist has offered no theory of the *phenomenal character* of experience—no theory of *what it's like* to see a mango, hallucinate the sound of a violin, or vividly imagine the scent of vanilla. In this chapter, I break that silence.

My focal point is the deep problem of whether or not perceptions and matching hallucinations have the same phenomenal character (section 10.1). To gather the resources to solve this problem, I take an extended detour into philosophy of language. I argue that the expression "what it's like" exhibits a particularly robust form of **context sensitivity** (sections 10.2–10.3), and when this expression has been used to define phenomenal character, at least three very different contexts have been operating in the background (sections 10.4–10.7). Thus, philosophers who speak of "phenomenal character" have had multiple properties in mind (section 10.8). This fractures the deep problem about "phenomenal character" into several shallow ones, which are easily solved (section 10.9).

10.1 A Deep Problem about Phenomenal Character

Consider two deeply attractive ideas. First, it seems obvious what it is like to *see* a scarlet plate is exactly the same as what it is like to *hallucinate* a scarlet plate. More generally, it seems obvious that:

PC1. Any perception and matching hallucination will have exactly the same phenomenal character.

Yet it is also attractive to think that the phenomenal character of any perception consists partly of the particulars that are perceived. For instance, when I see a scarlet plate, what my experience is like is constituted partly by *that very plate* and *its particular instances* of scarlet and roundness. *Those very particulars* are surely present in my stream of consciousness. But in a matching hallucination, there are no particulars to be experienced, so—apparently—its phenomenal character cannot be the same as that of the perception it matches:

PC2. It is not the case that any perception and matching hallucination will have exactly the same phenomenal character.[1]

That is the problem.

We can better understand the problem by seeing how it would bite against typical representationalist theories. A common representationalist view is that for any experience, its phenomenal character is identical to its property of being a perceptual representation with a certain content (or part of that content).[2] Given this assumption, we can generate a dilemma by asking, Is the phenomenal character of perception a matter of *at least partly singular* content or *purely nonsingular* content?

On the first horn, the representationalist answers that phenomenal character is *purely* a matter of *nonsingular* content. Then they can honor the claim that a perception and a matching hallucination will have the same phenomenal character: they can say that both experiences have exactly the same nonsingular content. But they lose their grip on the idea that particular objects and property-instances figure in the phenomenal character of perception. This claim is, I repeat, very plausible! It is very plausible that seeing a scarlet plate brings that very plate, and its particular instances of scarlet, into my phenomenal consciousness.

On the second horn, the representationalist instead answers that phenomenal character is a matter of content that is *at least partly singular*. This lets them honor the idea that my perception of a scarlet plate makes me phenomenally conscious of that very plate and its instances of scarlet, since they can say that these particulars are constituents of the content of my perception. However, on this view, no such particulars are constituents of the content of a matching hallucination; such an experience will have only a failed singular content. As a consequence, the representationalist is now forced to sacrifice the claim that a perception and a matching hallucination will have the same phenomenal character.[3]

This is one of the deepest problems for any theory of perception. Before we can solve it, we must take an extended detour into philosophy of language.

10.2 A Core Idea Introduced

A key part of my solution is the thought that different philosophers have used the term *phenomenal character* with importantly different properties in mind. In this section, I give a detailed analogy to characterize this core idea more precisely.

Begin by contrasting two sorts of context-sensitive expressions. On the one side, for a context-sensitive expression to be *shallow* is for it to have a reasonably narrow range of possible referents. A paradigmatic example is the context-sensitive expression *tall*. To be sure, in principle, this expression could refer to just about any height—just imagine that we are comparing mice. Nevertheless, *tall* will always refer to some height property or other. Similarly for other paradigmatic gradable adjectives: *rich, warm, funny*, and so on. Or, for a very different sort of shallow context-sensitive expression, consider *today*. This expression might in principle refer to any day, but it will always refer to some day or other.

On the other side, if a context-sensitive expression is not shallow, then it is *deep*. The expression *that* is an especially clear example of a deep context-sensitive term: in principle, it could be used to refer to nearly anything. I will also count slightly more specific expressions such as *that object* or *that property* as deeply context sensitive. For these expressions can in principle be used to refer to radically different objects or properties, respectively, so their range of possible referents is reasonably broad. By contrast, I will count expressions such as *that color* or *that number* as shallowly context sensitive, since there is some reasonably specific range of possible referents here.

The distinction between shallow and deep context-sensitive expressions is vague.[4] Still, a distinction does not have to be sharp to be real and useful.

Let us make one more distinction. Let *terms* be individuated by their meanings; let *expressions* be individuated by their phonemes and orthographies. Thus, there is a single expression, *bank*, corresponding to two different terms, *(river) bank* and *(financial) bank*. The expression *bank* has two different meanings, we might say.

Now return to philosophy of mind. By definition, the *phenomenal character* of an experience is what it's like for the subject to have that experience. I will soon argue that the defining expression *what it's like* is deeply context sensitive and that philosophers who discuss phenomenal character have had in mind different definitional contexts. But let me first give a less controversial example as a model.

Thomas N. points to a navy blue circle and says to his friends, "Let us refer to this property as *property X*." He is in a context in which color properties are salient. What can we say about Thomas's new term?

Thomas's definition relies on a deeply context-sensitive expression, *this property*. So, given Thomas's context, his term *property X* refers to navy blue. Surprisingly, Thomas's term *property X* is not context sensitive at all, *even though* it was defined using the deeply context-sensitive expression *this property*! The term *property X* simply refers to navy blue regardless of context.

Now suppose that, across the room, John C. is conversing with his friend Susanna S. and overhears Thomas. John's interest is piqued; he wants to talk about the same topic as Thomas. John points to the same navy blue circle and says to Susanna, "Let us refer to this property as *property X*." However, in John's context, what are salient are not colors but *conjoined color and shape properties*—for example, the property of being a navy blue circle. Still, John assumes that his term is the same as Thomas's and even intends for that to be the case when he gives his definition. Let us add, perhaps unrealistically, that John does not realize that the expression *this property* is deeply context sensitive. What can we say about John's term *property X*?

It is difficult to identify the referent of John's term, for there are two opposing pressures. On the one hand, John intends to be using the same term that Thomas is using. This tugs his term toward referring to navy blue. On the other hand, John has given an explicit definition of his term via the deeply context-sensitive expression *this property*. Given John's context, this tugs his term toward referring to the property of being a navy blue circle.

In light of these opposing pressures, I see three reasonable views about the referent of John's term *property X*. First, that it refers to navy blue; second, that it refers to the property of being a navy blue circle; and third, that it is indeterminate whether it refers to navy blue or to the property of being a navy blue circle. (I repeat that on any of these views, John's term *property X* will be context *in*sensitive.) I will not try to adjudicate among these views. I simply observe that whichever view is correct, there is an obvious sense in which the property of being a navy blue circle is what John *has in mind* when he talks about property X.

For example, when John discusses what he calls "property X," insofar as he is rational, what his evidence will *nonaccidentally support* are conclusions about the property of being a navy blue circle. He might say things like, "Necessarily, anything that has property X cannot be square or triangular." (His evidence might sometimes support conclusions about navy blue, but only by accident: "necessarily, anything that has property X is spatially extended.") Moreover, suppose that John were asked to rephrase his claims without using the expression *property X*. He would presumably use some expression referring to the property of being a navy blue circle.

It will be useful for us to develop our example a little further. Suppose that John and Susanna are not the only people who overhear Thomas. Many others do as well, and they too turn to their friends, point to the same navy blue circle, and define property X as "that property." However, they are speaking in still further contexts. For instance, in the context of Michael M., what are salient are conjoined shape and *determinable* color properties, such as the property of being a *blue* circle (rather than a *navy blue* one). Then there are Thomas's friends, such as David C., Frank J., and Michael T., who were conversing with Thomas and understood the original context. When all of these people mingle with one another and speak about "property X," they will have many different properties in mind. Mass confusion will be sown.

I say that something very similar has happened in contemporary philosophy of mind. Inspired by Thomas Nagel, philosophers have defined the *phenomenal character* of an experience as *what it's like* for the subject to have the experience.[5] But the defining expression "what it's like" is deeply context sensitive and has been used in the context of importantly different philosophical problems. At a minimum, these include the hard problems of consciousness, problems about a special kind of knowledge of reference, and problems about what experience makes directly available to the subject in a certain special way. Thus, philosophers who speak about "phenomenal character," whether or not they have been *referring* to the same property, have had importantly different properties *in mind*. These properties stand in interesting metaphysical relations to one another and are often co-instantiated. Thus, the slippage has not been easy to detect. But the properties *are* distinct. The result has been mass confusion.

The rest of this chapter defends this diagnosis and uses it to solve the deep problem about phenomenal character.

10.3 The Deep Context Sensitivity of "What It's Like"

According to the utterly standard definition, the *phenomenal character* of any experience is what it's like for the subject to have that experience. Equivalently, it is what *having* the experience is like for the subject; equivalently again, it is what *the experience* is like for the subject.[6]

Regarding that last equivalence: there is not *in general* an identity between what *something* is like for the subject and what *having that thing* is like for the subject. In the special case of experiences, however, it is plausible that there is such an identity: what having a perception of red is like for the subject is presumably the same as what that perception of red is like for the subject.

For us to understand the expression *phenomenal character*, we must first understand the expression *what it's like*. It is widely appreciated that the latter expression is context sensitive.[7] In this section, I argue that this is only the tip of the iceberg. What has remained below the water is that the expression is *deeply* context sensitive.

Suppose that, in making a claim about what something is like, some property *F* is cited. Sometimes, this will entail that the entity in question really is *F*: "I'll tell you what Sarah is like—she is honest." But sometimes, it will be obvious that there is no such entailment: "I'll tell you what blue is like—it is like purple." At other times, a property may not be cited at all: "I'll tell you what Chicago is like—it is like London."[8] To simplify the discussion, my proposal will focus only on uses of "what it's like" that are of the first type: *F-entailing uses*, as I shall say. I conjecture, but will not argue here, that the proposal generalizes to uses that are not F-entailing.

What I propose is:

> The **deep context-sensitivity view**: Any *F*-entailing claim of the form "what *x* is like is *F*" is deeply context sensitive, because it means that *F* is a property of *x* that is relevant in the context at hand (where this context may be further specified with prepositional phrases and subordinate clauses).

Is the deep context-sensitivity view best understood as a view about (i) *just* the literal meaning or (ii) the literal meaning *together with* the standard implicatures of claims of the form "what *x* is like is *F*"? This is a delicate question. Fortunately, the answer to it will not matter for my purposes. To glide over this question, I stipulate that I am using the term *meaning* to refer to the disjunction of (i) and (ii). Thus, the deep context-sensitivity view entails that an assertion that what x is like is F, where F is a contextually irrelevant property of x, is at least *inappropriate*, but it is neutral on whether such an assertion must also be *false*.

I offer seven lines of support for the deep context-sensitivity hypothesis.

The first line of support Suppose that we are discussing what India is like. On an internet travel forum, it might be appropriate to say, "It is full of temples, palaces, and other historical sites." In a geology class, it might be appropriate to say, "It has been largely shaped by extensive melting, and subsequent basalt flooding, underneath the Indian Craton." And in an economics conference, it might be appropriate to say, "It is enjoying rapid expansion—at present, it has the fastest-growing economy in the world."[9] Which claim is appropriate evidently depends on the context. It would be inappropriate, on the internet travel forum, to cite geological or economic features of India.

Thus, there is radical variation in which properties of India—including cultural properties, geological properties, and economic properties—can be appropriately cited as what India is like in different contexts. This directly supports the deep context-sensitivity view.

The same sort of evidence can be found when F-entailing claims about what something is like are adorned with a prepositional phrase or a subordinate clause. The traveler might discuss what India is like *in the winter*, the geologist might discuss what India was like *after the Deccan traps were created*, and the economist might discuss what India is like *in the manufacturing sector*. The deep context-sensitivity theorist should say that these adornments will serve to specify the context further (in addition to whatever role they play with respect to the literal meaning of the sentence).

The second line of support I have often tried to explain what I work on to students, friends, and other outsiders. But I have *never* been able to get *anyone* to understand the expression *phenomenal character* merely by giving the what-it's-like definition, with no prologue or epilogue. Even some professional philosophers find the term vague and unhelpful.[10]

This is puzzling. English speakers use what-it's-like talk all the time. How, then, can they fail to understand the what-it's-like definition of phenomenal character?

You might answer simply that what-it's-like talk is context sensitive. But this explanation is too weak. By analogy, imagine that someone wonders who the upper class are. You answer that they are the rich, with no prologue or epilogue. Even though the term *rich* is context sensitive, your audience will gain a good understanding of the term *upper class*. For the context sensitivity here is shallow: the term *rich* has a reasonably narrow range of possible referents.

However, if what-it's-like talk is *deeply* context sensitive, then the phenomenon is explained. For, on this view, saying that the phenomenal character of an experience is what it's like for the subject to have that experience, with no prologue or epilogue, is much like saying, "The phenomenal character of an experience is *that property*," or, "The upper class are the individuals with *that property*," with no hint of which property you are referring to. Your audience will be mystified.

The third line of support I tell my introductory students to read Nagel's classic paper "What Is It Like to Be a Bat?" which offers a what-it's-like definition of the term *subjective character*. They understand this—but they are baffled when the same definition is served neat. How is this possible?

The deep context-sensitivity view provides a satisfying answer: Nagel's article provides the missing context for the what-it's-like definition. To return to our original example, it is unhelpful to hear, out of context, that property X is *that property* of a navy blue circle. But the very same remark would be very informative once you learn that the discussion is about color properties.

The fourth line of support It is often said that there is some deep connection between the what-it's-like properties of experience and consciousness. It is also often noted that this is semantically mysterious.[11] Look at the words "what," "it's," and "like." Individually, they have no particular connection to consciousness, nor does putting them together seem to help. What is happening?

Some theorists propose complex answers to this question.[12] I offer a simple one: *out of context*, there is no connection between the expression *what it's like* and consciousness. The connection arises almost entirely from the context itself. By analogy, out of context, there is no particular connection between the expression *that* and the color navy blue. But it is unmysterious how such a connection might arise in certain contexts.

The fifth line of support It is sometimes held that philosophers are using the expression *what it's like* in a special, technical sense.[13] This would explain why the bare what-it's-like definition is unhelpful. It might also explain why Nagel's article is helpful: perhaps students can glean this special, technical sense of the expression *what it's like* from context. It might also explain why examining the words "what," "it's," and "like" reveals no special connection to consciousness.

Better, though, not to posit special technical senses without necessity. That is a virtue of the deep context-sensitivity hypothesis: it does justice to the data while saying that philosophers use the expression *what it's like* in a perfectly ordinary way.[14]

The sixth line of support Suppose you ask what it was like for me to see Cirque du Soleil in New York. I might truly answer, "Surprising and exciting!" Thus:

VPC1. What it's like for me to visually experience Cirque du Soleil is surprising and exciting.

("VPC" stands for "visual phenomenal character.")

However, philosophers usually agree that the properties of being surprising and exciting are not *phenomenal characters* of the experience of seeing Cirque du Soleil. To be sure, philosophers *do* usually think that the events of surprise and excitement *have* phenomenal characters of their own. But they do not think that the properties of being surprising and exciting *are themselves* phenomenal characters of my visual experience of Cirque du Soleil. For example, various philosophers have identified the total phenomenal character of visual experiences with—for example—(i) the property of being a special type of representation of certain perceived entities; (ii) the property of being a primitive, nonrepresentational relation of awareness to certain perceived entities; and (iii) the neural property that underlies my visual perception.[15] None of these theories would treat the properties of being surprising and exciting as phenomenal characters of my visual experience of Cirque du Soleil.[16]

Thus:

VPC2. The properties of being surprising and exciting are not phenomenal characters of my visual experience of Cirque du Soleil.

But, by definition:

VPC3. The phenomenal character of my visual experience of Cirque du Soleil is what it's like for me to visually experience Cirque du Soleil.

These claims seem contradictory! What is going on?

The deep context-sensitivity view gives us a satisfying answer to this question. Recall Thomas N., who, speaking in a context where color properties are salient, points to a navy blue circle and says, "Let us refer to this property as *property X*." The term *property X* refers to navy blue. Still, because the expression "this property" is deeply context sensitive, we can easily use it to refer to properties that are unrelated to navy blue.

For the same reason, given the deep context-sensitivity view, VPC1–VPC3 do not contradict each other. It is just that the expression "what it's like to have an experience" is deeply context sensitive. We use that expression to refer to one set of properties—phenomenal characters—in VPC3, while referring to a very different set of properties—emotional properties—in VPC1.

The seventh line of support Philosophers have developed many theories of the meaning of "what it's like." Almost all of these support the deep context-sensitivity view.

For example, consider:

> The *resemblance theory*: Any claim of the form "What *x* is like is . . ." literally means, "What *x* resembles is . . ." or, equivalently, "What *x* is similar to is. . . ."[17]

The resemblance theorist offers a (more or less) unified semantic analysis of sentences such as the following:

(1) What Chicago is like is London.

(2) What Chicago is like is exciting.

The analysis of (1) is obvious enough: this sentence means that Chicago *resembles* London (i.e., that Chicago is *similar to* London). The analysis of (2) is what is interesting. This does not mean that Chicago is similar to exciting; that makes no sense. Rather, it means that Chicago is similar to *things* that are exciting, *in respect of being exciting*. Thus, (2) has the same truth-conditions as:

(3) Chicago is exciting.

But (2) and (3) are quite different in semantic structure.[18]

The resemblance theorist should welcome the deep context-sensitivity view with open arms. For resemblance-talk (i.e., similarity-talk) is deeply context-sensitive. For example, different claims of the form "India is similar to . . ." will be appropriate in different contexts: the relevant similarities might be cultural, geological, economic, and so on.

Now consider this theory:

> The *predicative theory*: Any *F*-entailing claim of the form "what *x* is like is *F*" literally means that *x* has property *F*.

Some predicative theorists hold that the expression "what *x* is like is *F*" gets its literal meaning idiomatically.[19] Others say that it gets its literal meaning compositionally—for example, that it means, "Experience *E* is similar to experiences that are *F* in respect of being *F*—that is, *E* is *F*." (Notice that this is a way of combining the predicative theory with the resemblance theory.)[20] Still other predicative theorists have remained neutral between the idiomatic and compositional approaches.[21]

There is also a close cousin of the predicative theory, namely:

> The *restricted predicative theory*: Any *F*-entailing claim of the form "What *x* is like is *F*" literally means that *F* is a way that *x* is.[22]

Ways are to be understood as properties of a special type: those that can be cited in response to questions about *how* something is, as opposed to *where* it is or *why* it is.[23]

The predicative and restricted predicative theories would both support the deep context-sensitivity view. For, as I hope is evident by now, there is deep context sensitivity in claims about (i) what properties India has and (ii) how India is.

The deep context-sensitivity view would be equally supported by the following theory, which is an extension of the restricted predicative theory:

> The *affective theory*: Any *F*-entailing claim of the form "What it is like to *x* for *y* to φ is *F*," when uttered in context *c*, literally means that *F* is a way that *y*'s φ-ing affects *x* in *c*.[24]

Consider claims about what is it like, to you, for you to have a dog. The affective theorist would understand these as claims about the ways in which your having a dog affects you. If we are discussing your physical condition, then you might say, "Having a dog affects me by being helpful—in particular, encouraging me to get a lot of exercise"; if we are discussing your mental condition, then you might say, "Having a dog affects me by being deeply meaningful"; if we are discussing your financial condition, then you might say, "Having a dog affects me by draining my bank account!" These properties—the properties of being helpful, being deeply meaningful, and draining one's bank account—are properties of radically different types.

Finally, consider:

> The *perceptual appearance theory*: Any claim of the form "what experience *E* is like for subject *S* is *F*," as used in philosophy of mind, literally means that *F* is a way that *E* perceptually appears to *S*.[25]

This theory is supposed to entail an inner-sense theory of phenomenal character—the idea being that an inner-sense theory is *analytically* true.[26]

If the perceptual appearance theory is true, then the deep context-sensitivity view, even if it is true, is irrelevant to philosophy of mind. For the deep context-sensitivity view is a view about *F*-entailing claims of a certain form, and the perceptual appearance theory says that there are no such *F*-entailing claims in philosophy of mind: the fact that an experience *perceptually appears* to be *F* does not entail that it *really is F*.

That said, I believe that there are serious problems with the perceptual appearance theory. Many of these criticisms have been detailed by others.[27] In addition, this theory does not enjoy the first, second, third, fourth, or sixth lines of support described in this section. Finally, I do not see how the theory can adequately account for the importantly different uses of the term *phenomenal character* that I will point out in the remainder of this chapter.

I conclude that the deep context-sensitivity view receives strong support from almost every theory of the literal meaning of what-it's-like talk. The major exception is the perceptual appearance theory. While this theory does not undermine the deep context-sensitivity view, it does render it trivial. However, I believe that there are strong reasons to reject the perceptual appearance theory anyway.

Conclusion The deep context-sensitivity view enjoys the following lines of support:

(a) It is directly supported by ordinary examples of what-it's-like talk.
(b) It explains why the bare what-it's-like definition of phenomenal character is mystifying.
(c) It explains why the what-it's-like definition becomes helpful when given in a clear philosophical context of the sort that Nagel provides.
(d) It explains how it is possible for the phenomenal character to be deeply linked to consciousness, even though the words "what," "it's," and "like" have no particular relationship to consciousness.
(e) It treats philosophical uses of *what it's like* as entirely ordinary rather than technical.

(f) It explains why, in nonphilosophical contexts, we can talk about what an experience is like without talking about its phenomenal character.

(g) It is supported by all of the best theories of the meaning of "what it's like."

I find this to be compelling evidence for the deep context-sensitivity view.

Looking forward If the deep context-sensitivity view is true, then, to understand the meaning of the term *phenomenal character*, we must first identify the philosophical context in which this term is defined. I will argue that philosophers have, without realizing it, used the what-it's-like definition of *phenomenal character* with at least three different definitional contexts in mind (sections 10.4–10.7).

10.4 A Failed Attempt at Supplying Context

There are certain standard supplementary remarks that philosophers tend to make when they introduce the term *phenomenal character*. Might these remarks suffice to pin down the relevant philosophical context(s)?

One standard tactic is to mention a few examples: to say that there is something it's like to see purple, to hear a cello, and so on.[28] However, this popular supplement provides very little of the missing information about the context(s) that philosophers have in mind. For notice: if you ask someone what it's like to see purple, they may truly answer, "It's simply *wonderful*." But, again, it is widely agreed that emotional states or properties are not themselves phenomenal characters of visual experiences.

Another common way to fill in some missing contextual information is to connect the term *phenomenal character* to other terms in the neighborhood: perhaps *subjectivity*, *awareness*, or *consciousness*.[29] But this is not sufficient. For one thing, these terms are wildly polysemous—for instance, it has been said that the term *consciousness* has more than half a dozen distinct meanings.[30] Regardless, these terms do not supply enough of the missing contextual information. If you ask someone what it's like to have a particular episode of *subjective conscious awareness*, then they can truly answer by reporting their emotions.

Thus, the standard supplementary remarks do not, by themselves, pin down the meaning of the term *phenomenal character*. Philosophers have something more in mind.

What might that be?

10.5 The First Context: Hard Character

I observe that in one strand of the philosophical literature, the term *phenomenal character* gets used in the context of the so-called *hard problem(s) of consciousness*. In particular, at least in paradigmatic cases, these theorists take it that the phenomenal character of an experience plays the following roles:

1. The *explanatory gap role*: We seem unable to understand how it could be nothing more than something that is conceived purely physically and/or functionally. (See Levine 1983.)
2. The *missing knowledge role*: We seem unable to know everything about it just on the basis of having all physically and functionally conceived information. (See Nagel 1974; Jackson 1982.)
3. The *conceivable absence role*: We seem to be able to conceive of it as present in one subject *S* yet absent in some other subject who is identical to *S* in all functional and physical respects. (See Block 1980; Chalmers 1996.)
4. The *conceivable inversion role*: We seem to be able to conceive of it as invertible across subjects who are identical in all functional and physical respects. (See Shoemaker 1982.)[31]

Why think that the context characterized by roles 1–4 pins down one property that philosophers have had in mind when using the expression *phenomenal character*? One compelling reason is sheer influence. Look at discussions of phenomenal character from the final decades of the twentieth century. You will find over and over that it is the problems implicated in roles 1–4 that are discussed, and it is the works listed above that are cited.

In addition, in the more recent literature, many philosophers assert that there is some especially close connection between phenomenal character and roles such as 1–4, especially the explanatory gap role. Below are a few representative quotations. Note that although these quotations use terms other than *phenomenal character*, these other terms are always defined with the help of the expression *what it's like*:

> A property is phenomenal only if it contributes to the hard problem of consciousness, and in particular, only if it gives rise to an explanatory gap. (Carruthers and Veillet 2011, 45)

> Perhaps the most striking feature of phenomenal properties is that they appear to present an explanatory gap. (McClelland 2016, 540)

> The traditional explanatory gap consists in the fact that phenomenal states do not seem to be explainable in terms of physical states [see Levine 1983]. Phenomenal states are states that there is *something it is like to be in*. In what follows, I distinguish four features of phenomenal states that individually resist explanation." (Saad 2016, 2356; reference to Levine is his)

> *Consciousness properties* are properties that contribute to making up what things are or can be *like* for subjects. . . . I will often abbreviate the physicalist identity thesis as the claim that C=P. . . . Many of us have an exceptional resistance to the physicalist identity thesis. . . . Our exceptional resistance to physicalism is the "explanatory gap" on one understanding of that phrase. (Sundström 2018, 681–682)

But why think that these remarks are intended as *reference-determining* contextual information for the expression *what it's like*, when that expression is being used to define the term *phenomenal character*? Couldn't these simply be further assertions about phenomenal character?

I doubt it. It is very clear that all of these theorists have in mind *some* restrictions when they use the term *phenomenal character*. None of them would count *every* property of an experience as a phenomenal character of that experience. But I cannot find anything else in these works that would supply these contextual restrictions. The remarks that I have quoted are the only substantial, dialectically neutral bits of context that these theorists give before they launch into their arguments.

In other cases, I would offer the same interpretation on more holistic grounds. A paradigmatic example is Michael Tye's 1995 book *Ten Problems of Consciousness*. In the first paragraph of chapter 1 of this book, Tye simply lists a few experiences, mentions their "phenomenal or 'what it is like' aspects," and moves on. But it is impossible to miss the imprint of the hard problems on Tye's way of thinking: the titular ten problems include the explanatory gap problem, the missing knowledge problem, the conceivable absence problem, and the conceivable inversion problem.[32] Similarly for Dretske's book *Naturalizing the Mind*. Dretske briskly defines the "subjectivity" of an experience of *F* as "what it is like to experience *F*."[33] Immediately afterward, however, he launches into a fifteen-page discussion of the missing knowledge role, as characterized by Jackson and Nagel.[34]

There is a final reason for thinking that roles 1–4 help to fix what some philosophers have in mind when using the term *phenomenal character*. Recall my earlier report that I have *never* been able to get the uninitiated to understand the term just by appealing to what it's like to have an experience. I now add that I have *routinely* been able to induce understanding just by describing roles 1–4—and that has been the *only* way that I have found to get the uninitiated to use the term *phenomenal character* in the way that these philosophers do. It has made no difference whether I use the expression "what it's like."

I conclude that when these philosophers use the expression "phenomenal character," it is *hard character* that they have in mind. As a reminder, here is the definition of hard character from chapter 4:

> If there is a single natural kind property of experience that tends to generate enough of problems 1–4 in paradigmatic cases, then the *hard character* of a mental item is the item's property that belongs to this natural kind. If there is no natural kind that generates enough of problems 1–4, then *hard character* simply refers to any property of that experience that generates enough of these problems. If there several natural kinds that generate enough of problems 1–4, then *hard character* refers indeterminately to these.

This is precisely what these philosophers wish to discuss.

10.6 The Second Context: Strong Referential Character

Next, turn to philosophers working, not in the context of the *hard problems of consciousness*, but in the context of certain problems about *perception*—the paradigm being John Campbell in his 2002 book *Reference and Consciousness*. It is a central aim of Campbell's book to account for the "phenomenal character" of perception. Yet Campbell has little interest in the hard problems of consciousness: his main discussion of phenomenal character in chapters 6 and 7 contains not a single mention of explanatory gaps, inverted spectra, or zombies. These phrases do not even appear in the book's index.[35] What is going on?

I suggest that when Campbell uses the term *phenomenal character*, what he has in mind is not hard character but *strong referential character*. This, you will recall from chapter 7, is defined as the property of an experience that explains why it positions the subject to make knowledgeable *de re* reference to its targets.

Let me build my case for this interpretation gradually. My focus will be on chapter 6 of *Reference and Consciousness*, for it is here that Campbell carefully introduces the notion of phenomenal character and offers his theory of it. At almost the very beginning of that chapter, Campbell makes some remarks that are worth quoting at length:

> Suppose that you live in one of a terraced row of houses, and you sometimes hear noises from the house next door. Being of an enquiring turn of mind, you formulate hypotheses about what objects are to be found next door. . . . And you might test and confirm your hypotheses over a long period, without ever catching sight of those things.
>
> Suppose now that the day finally arrives when you do get a look inside the house. What does this add to your knowledge? Perhaps the hypotheses you formed had been amply confirmed long before your look inside, so the existence of objects with these particular functional roles does not get significant further confirmation from your observation. Nor is it that you can now refer to those particular objects but could not refer to them before. You could have referred to those particulars before. . . .
>
> The contrast between the knowledge you have now, on the basis of a look at the objects, and the knowledge you had before of the existence of objects with particular functional roles, is that when you see the thing, you are confronted by the individual substance itself. . . .
>
> These remarks fill out something of the sense in which conscious attention to an object can provide you with knowledge of the reference of a demonstrative referring to it. (2002, 115–116)

This is Campbell's characterization of a special kind of knowledge of reference. Note that this kind of knowledge of reference is associated distinctively (though, as I argued in chapter 7, not exclusively) with perception. *Seeing* a mahogany table will position you to make knowledgeable direct reference to it. Merely *hallucinating* a table, or *sensorily imagining* a table that you have not seen before, will not even position you to refer to a real table, much less to know what you are thereby referring to.

A few pages later, Campbell makes the following telling remark:

> Since both the Relational and Representational Views of the phenomenal character of experience have to be taken seriously . . . , how are we to decide between them? . . . I think that the only way to proceed is to ask why we need the notion of the phenomenal character of experience. We have to look at the role that the notion plays in our reflective thinking, we have to ask what the point is of the notion. I have been arguing that experience is what explains our grasp of substantial objects. (Campbell 2002, 120)

Given Campbell's emphasis on the *notion* of phenomenal character, it is clear that this remark is intended to be definitional. To be sure, at the end of the quotation, Campbell slides from talk about the *phenomenal character* of an experience to talk of *the experience itself*. But the term *phenomenal character* does not appear anywhere else in the paragraph from which this quotation was taken. Thus, the charitable interpretation is that, according to Campbell, the point of the notion of phenomenal character is to explain our grasp of substantial objects.

It is also clear that Campbell's talk of "explaining our grasp of substantial objects" is meant to refer to special knowledge of reference. For one thing, Campbell goes on to argue that the special sense in which we can know the reference of perceptual demonstratives is partly a matter of our being able to form a certain conception of the objects being demonstrated—for instance, a conception of those objects as having categorical and intrinsic characters. It is in this sense that we have a grasp of objects as *substantial*. In addition, the present quotation is from a section entitled "The Argument from the Explanatory Role of Experience," and the section from which the earlier quotation was drawn is entitled "The Explanatory Role of Experience." Evidently, Campbell intends for his argument here to draw on the explanatory role that he identified earlier.

Further support for this interpretation can be found in chapter 7, where Campbell says,

> Why do we need the notion of the phenomenal character of experience? We have to look at the role that the notion plays in our reflective thinking, we have to ask what the point is of the notion . . . I have argued that we use the notion of experience of an object—or more precisely, conscious attention to an object—in explaining how it is that we have knowledge of the reference of a demonstrative. (Campbell 2002, 138)[36]

Here Campbell makes it explicit that he is discussing special knowledge of reference. Again, Campbell's use of the word *notion* makes it clear that his remark is intended to be definitional.[37]

Thus, I conclude that in his book *Reference and Consciousness*, Campbell uses the term *phenomenal character* with strong referential character in mind. For reasons explained in section 10.2, this conclusion is silent about the *referent* of Campbell's expression. If his dominant intention is to use the same term as certain previous thinkers, then his expression may *refer* to hard character, even while strong referential character is what he *has in mind*.

10.7 The Third Context: Strong Accessible Character

That is not all; there is yet a third context in play.

Consider Susanna Siegel's 2010 book *The Contents of Visual Experience*, which theorizes heavily about "phenomenal character." Siegel is not terribly explicit about how she is using that expression. Like most authors, Siegel introduces phenomenal character simply by giving the "what-it's-like" definition, along with an example of an experience. This does not give us much information.

However, Siegel does leave behind some interesting clues. To begin with, Siegel presupposes that if there are properties presented in *phenomenology*—which she takes to be equivalent to *phenomenal character*[38]—then "those properties are conveyed to the subject."[39] Now, from context, it is clear that this is presupposition is meant to be acceptable to all parties in the dispute. This suggests that Siegel is thinking of phenomenal character as something like the property of conveying . . . to the subject.

In a similar vein, Siegel describes one of her arguments, the *Argument from Appearing*, as "starting with something that is conveyed to the subject."[40] Shortly afterward, Siegel characterizes this same argument as "proceed[ing] from premises about the phenomenal character of visual perceptual experience." This further supports the idea that, for Siegel, the phenomenal character of an experience is its property of conveying . . . to the subject.

But what is it for an experience to *convey* something? Siegel clarifies that an experience conveys a content "if [that content] would be a content of explicit beliefs that are natural to form on the basis of visual experience," "if [the content] enables the experience to guide bodily actions," and "if it is manifest to introspection that [the content] is a content of experience."[41]

In light of this, consider this definition:

> The **weak accessible character** of a mental item is its maximal property of the form, *making . . . directly available to its subject*. Thus, what it is for a mental item to have accessible character—equivalently, to be *weakly access-conscious*—is for it to make *something* directly available to its subject.

Here, when I speak of a *maximal property*, I mean that any mental item has exactly one, all-inclusive weak accessible character—a single property that covers *everything* that the mental item makes directly accessible to its subject. In addition, what I am calling *weak access consciousness* is usually just called *access consciousness*.[42] I am using the more complex terminology because I will later contrast this with *strong* access consciousness.

Siegel's remarks suggest that she is defining phenomenal character as weak accessible character. For the contents implicated in weak accessible character will be directly available in precisely Siegel's sense: they will influence beliefs, guide actions, and be accessible to introspection.

However, this interpretation cannot be quite right: Siegel takes it to be an open question "whether there can be phenomenal consciousness [the kind of consciousness characterized by phenomenal character] without what Ned Block calls 'access consciousness'" (2010, 81, fn. 3). So Siegel must be understanding phenomenal character in some other way. What could that be?

To set up my answer, I offer a few more definitions:

> If a mental item is hard-conscious, then its **strong accessible character** is its maximal property of the form, *making . . . directly available to its subject.* If a mental item is not hard-conscious, then it has no strong accessible character. Thus, what it is for a mental item to have strong accessible character—equivalently, to be *strongly access-conscious*—is for it to be hard-conscious and to make something directly available to its subject.

The idea is that hard consciousness is a *background condition* for strong accessible character. By analogy, some have thought it plausible that pieces of wood arranged table-wise constitute a table *only if* they were assembled by someone with the right intentions, that particles arranged human-wise constitute a human being *only if* they have the right origins, and that an object constitutes a work of art *only if* it is treated in the right way by some community. These are all examples of background conditions. Similarly, I am stipulating that there are background conditions for strong accessible character: the property of making . . . directly available to the subject constitutes strong accessible character *only if* the mental item is hard-conscious.

You might wonder whether I have really identified a third kind of consciousness/character here. For it is usually thought that every mental item that is hard-conscious is access-conscious. Thus, won't the notion of strong access consciousness/character be identical to hard consciousness/character?

I grant that these properties are necessarily coextensive: as a matter of metaphysical necessity, something is strongly access-conscious *just in case* it is hard-conscious. But I say that these properties are nevertheless distinct.

To appreciate how this is possible, suppose that, necessarily, something has mass just in case it has spatiotemporal location.[43] Then compare the properties of (i) *having mass and spatiotemporal location* and (ii) *having mass.* An entity will have *some* mass-and-location property just in case it has *some* mass property, but *which* specific mass-and-location property it has will be different from *which* specific mass property it has. Thus, (i) and (ii) are very different properties. Similarly for the relationship between strong accessible character and hard character. A mental item will have *some* strong accessible character and will thereby be strongly access-conscious, just in case it has *some* hard character and is thereby hard-conscious. However, *which* specific strong accessible character it has will differ from *which* specific hard character it has. Thus, strong accessible character/consciousness is distinct from hard character/consciousness.

Suppose, for example, that I see a scarlet pepper under ordinary circumstances. According to the rich pluralist, the hard character of my experience might simply be the property of being an instance of deep awareness of scarlet and pepper-shapedness. By contrast, the rich pluralist will take the strong accessible character of my experience to be much more complex. It will be something like the property of making all the following directly available to the subject:

(i) the universals scarlet and pepper-shapedness,
(ii) the particular pepper and its particular instances of scarlet and pepper-shapedness, and
(iii) the conceptualized contents of being red[44] and being a pepper.

The rich pluralist says that is made directly available to the subject (i) by my deep awareness, (ii) by my sensory representations, and (iii) by my categorization of the pepper *as* red and *as* a pepper. Interestingly, this is the same as the weak accessible character of the experience. More generally, *for any hard-conscious mental item*, its weak accessible character will be the same as its strong accessible character. But weak accessible character and strong accessible character are not the same in general: a hard-unconscious mental item can have weak accessible character, but it cannot have strong accessible character.

I offer a brief aside. Some have argued that it is most innocent, and thus best, to define the term *phenomenal character* simply by giving a long list of positive and negative examples.[45] However, I submit that this approach promotes ambiguity. It cannot tell us whether *phenomenal character* refers to hard character or to strong accessible character: I have argued that these properties are distinct but necessarily occur in exactly the same cases. That is why I have defined different kinds of "phenomenal character" by appealing to examples *together with* theoretical roles.

I now offer my final interpretation of Siegel: I think that she means to define *phenomenal character* as *strong accessible character*. This would explain why she thinks that phenomenal character, by definition, concerns what is *conveyed* to the subject: it is because, by definition, strong accessible character concerns what is *made directly available* to the subject. This interpretation would also explain why Siegel regards it as an open question whether phenomenal character is the same as weak accessible character: it is because the definition of strong accessible character guarantees that it will always *co-occur* with weak accessible character, without settling whether these properties are *identical*.

10.8 The Fragmentation of "Phenomenal Character"

I have argued that when discussing "phenomenal character," philosophers have had at least three different definitions in mind, namely:

> The phenomenal character of any mental item is its *hard character*—roughly, the property that generates the hard problems of consciousness. (The precise definition is more complex and was given earlier.)
>
> The phenomenal character of any mental item is its *strong referential character*: its maximal property of the form, *positioning the subject to make knowledgeable reference to. . . .*
>
> The phenomenal character of any mental item is its *strong accessible character*: its maximal property of the form, *making . . . directly available to its subject.*

It is evident that these *definitions* are all distinct. In this section, I will argue that the definitions pick out at least two distinct *referents*.

It is clear that hard character must be distinct from strong referential character and strong accessible character. When I see a scarlet pepper, the particular pepper will not be a constituent of the hard character of my experience. As I argued in chapter 4, hard character is not constituted by external particulars; it is constituted by partly revelatory awareness of the sensory qualities alone. However, the particular pepper *will* be a constituent of the strong referential character and the strong accessible character of my experience. For my experience positions me to make knowledgeable reference to the pepper (see chapter 7), and it makes the pepper itself directly available to me in a hard-conscious way.

As for strong referential character and strong accessible character, I suspect that they are the same. Both require hard consciousness (deep awareness), and both are constituted by sensory qualities, the instances of those qualities, and the particulars that have those property-instances. But perhaps there are exotic cases in which these two kinds of character come apart.

I conclude that philosophers who use the term "phenomenal character" have had in mind not only at least three different *definitions* but also at least two different *referents*.

An objection Some readers might continue to think that everyone—including Tye, Dretske, Campbell, and Siegel—has had a single property in mind when discussing phenomenal character. The thought is that these thinkers introduce the same term in very different ways because they are highlighting *synthetic* rather than *analytic* truths about phenomenal character—but they all have the same analytic truths in mind.

A suggestion like this was made by no fewer than five people who read a draft of this book. What I find interesting is how these readers spelled out the concern.

Two of these readers, Boyd Millar and Brian Cutter, were representationalists. They took all parties to agree that there was an analytic connection between phenomenal character and the hard problems of consciousness. In other words, they thought that everyone had in mind the concept of *hard character*. On their interpretation, Campbell was highlighting the synthetic fact that hard character is a necessary condition for strong referential character, and Siegel was doing the same for the synthetic fact that hard character is made directly available to the subject.

My third and fourth readers, Ivan Ivanov and Umrao Sethi, were naïve realists. They were confident that it was the concept of *strong accessible character* that all parties had in mind. They thought that Nagel, Tye, and company were simply noticing that strong accessible character synthetically generates the hard problems of consciousness.[46] Likewise, they read Campbell as making the synthetic observation that the strong accessible character *of perception* positions the subject to know, in a special sense, what they are referring to.

My fifth reader was Heather Logue, a hybrid representationalist-cum-naïve-realist. She gave yet another interpretation. She thought that *all* of the connections mentioned above—between phenomenal character and (i) the hard problems, (ii) knowledge of reference, and (iii) what was directly available to the subject—were synthetic. She assumed that all parties to the dispute accepted this. She instead took it as analytic that phenomenal character constitutes the subject's conscious perspective on the world, where the relevant notion of consciousness was pinned down with a long list of positive and negative examples.

In short, once the different concepts of "phenomenal character" were distinguished, these five readers agreed that everyone has had a single concept in mind all along. *But they massively disagreed about which concept that was.* Far from constituting an objection, then, these results reinforce my conclusion: philosophers who discuss "phenomenal character" have had different definitions, and different referents, in mind.

With this result in hand, we are ready to return to the deep problem about phenomenal character.

10.9 The Deep Problem Dissolved

The deep problem consists of these plausible but inconsistent claims:

PC1. Any perception and matching hallucination will have exactly the same phenomenal character.

PC2. It is not the case that any perception and matching hallucination will have exactly the same phenomenal character.

As we now know, there are at least two, and perhaps three, different kinds of "phenomenal character." To bring the problem into focus, we must clarify precisely which kind is at issue: hard character, strong referential character, or strong accessible character?

Perhaps the problem is supposed to be about *hard character*. Then the claims at issue are these:

HC1. Any perception and matching hallucination will have exactly the same hard character.

HC2. It is not the case that any perception and matching hallucination will have exactly the same hard character.

How should we respond to this problem?

As I argued in chapter 4, the hard character of any experience—the natural kind property that generates the hard problems—is simply its property of the form, *providing partly essence-revealing awareness of* ___. The blank is to be filled in with sensory qualities alone, for, as I argued in chapters 2–4, only these have their essences partly revealed to us in experience. I also argued that partly essence-revealing awareness will be shared across any perception and matching hallucination. These experiences might make us aware of different *external particulars*, but that is irrelevant: such awareness is not partly revelatory. What matters is that any perception and matching hallucination will partly reveal the essences of exactly the same *sensory qualities*.

Thus, we should reject HC2. This solution comes at little cost, since it is clear that awareness of external particulars plays no role in generating the hard problems. To give just one reminder of the point, recall:

CQ4. Why do you experience *this* pepper rather than experiencing an intrinsically identical pepper somewhere in Vietnam?

This is not a hard problem of consciousness, because it only concerns awareness of different external particulars.

I conclude that it is easy to solve *this* version of the deep problem about phenomenal character.

But suppose that we understand phenomenal character as strong referential character. Then the deep problem about phenomenal character becomes this:

SRC1. Any perception and matching hallucination will have exactly the same strong referential character.

SRC2. It is not the case that any perception and matching hallucination will have exactly the same strong referential character.

This time, we should reject the first claim and preserve the second one. As I argued in chapter 7, SRC1 is not even prima facie plausible. Seeing a pepper positions the subject to make knowledgeable reference to certain sensory qualities *and* to a particular pepper; hallucinating a pepper positions the subject to make knowledgeable reference to the same sensory qualities but not to any particular pepper. (There is no pepper there, so how could the subject refer to it in the first place, much less know what they are thereby referring to?) Again the problem melts away.

Finally, suppose that we focus on strong accessible character. Then the problem becomes this:

SAC1. Any perception and matching hallucination will have exactly the same strong accessible character.

SAC2. It is not the case that any perception and matching hallucination will have exactly the same strong accessible character.

I think that SAC1 is obviously false. When I see a scarlet pepper, I have a hard-conscious experience that makes me aware of *that particular pepper.* When I have a matching hallucination, my experience is still hard-conscious, but it does not make me directly aware of any particular pepper. Thus, these experiences clearly differ in their strong accessible character.

To be sure, in one important sense, the perception and the hallucination will *introspectively seem* to have the same strong accessible character. In particular, take any claim expressed by a sentence of the form "I introspectively seem to be experiencing ___," where the blank is filled with a purely existential description (e.g., "a mind-independent scarlet pepper"). That claim will be true of my perception just in case it is true of my hallucination. The rich pluralist's theory of apparent presentational character, given in chapter 8, explains why this is so.

I conclude that this third version of the deep problem is also easily solved.

10.10 Conclusion

I have argued that philosophers have unwittingly used the expression *phenomenal character* while having in mind at least three different definitions, which refer to at least two different properties. These properties stand in interesting metaphysical relations to one another, in something like the way that there are interesting metaphysical relations between being navy blue and being a navy blue circle. Still, the properties are distinct. Once we realize this, the deep problem about "phenomenal character" fractures into several different but quite shallow problems, which the rich pluralist can easily solve. These solutions come at little cost: the rich pluralist can adopt them while paying full respect to *all* of the motivating intuitions about each kind of phenomenal character.

It is not clear to me that this advantage is distinctive to rich pluralism. I suspect that disambiguating the expression "phenomenal character" makes it possible for many different theorists to solve the deep problem. That is why this chapter is in part IV of the book: I am not claiming to find new comparative advantages for rich pluralism, but just trying to solve whatever problems remain.

There is only one cluster of problems left: the rich pluralist must confront a series of arguments from hallucination. That is the topic of the next chapter.

11 Arguments from Hallucination

There is not just one argument from hallucination; there are many. In this chapter, I consider three such arguments: a basic argument from hallucination (section 11.1), a screening-off argument (section 11.2), and an argument from introspective access (section 11.3). I rebut each one on behalf of the rich pluralist.

Other theorists may be able to use these same rebuttals or equally good ones. However, I will let my opponents fend for themselves. My goal is purely defensive: I aim to show that rich pluralism, at least, can provide satisfying responses to these arguments from hallucination.

11.1 A Basic Argument from Hallucination

Compare two qualitatively identical subjects, only one of whom is in genuine perceptual contact with their environment. For the sake of concreteness, let us suppose that the first subject enjoys a veridical perception of a tomato, call it the *tomato-perception*, which involves a certain neural event. The second subject suffers a matching hallucination, call it the *tomato-hallucination*, which involves a qualitatively identical neural event but is caused by the manipulations of a devious neuroscientist who ensures that its proximal cause is qualitatively identical to the proximal cause of the tomato-perception.

With this case in mind, consider:

The *basic argument from hallucination*:

A1. The tomato-hallucination does not present the subject with a tomato.

A2. If the tomato-hallucination does not present the subject with a tomato, then the tomato-perception does not present the subject with a tomato.

A3. The tomato-perception does not present the subject with a tomato.

A1 is extremely plausible, since (we may stipulate) there is no tomato for the hallucinating subject to be presented with. A3 is inconsistent with rich pluralism. Thus, the rich pluralist should reject A2.

You might try to support A2 by observing that the neural events underlying the perception and the hallucination are effects of qualitatively identical proximal causes on qualitatively identical subjects. However, even assuming that qualitatively identical causes will have qualitatively identical effects, this shows only that these neural events are qualitatively identical. It does not show that these events must constitute events of the same *mental kind(s)*. It is entirely possible that the tomato-hallucination is constituted only by the neural event and thus does not present a tomato, while the tomato-perception is constituted not only by the neural event but also by a relation (say, one of successful sensory representation) to a real tomato and partly for that reason *presents* a tomato. Lo and behold, this is the rich pluralist's view.

But there is another line of support for A2: the tomato-hallucination introspectively seems to be the same as the tomato-perception in all qualitative respects. This does not *entail* that these experiences are qualitatively identical, since two mental events can be radically qualitatively different even if they introspectively seem to be the same. But we need some kind of explanation for the introspective appearance of sameness.

The categorization account gives the rich pluralist such an explanation: both experiences are constituted by deep awareness of precisely the same sensory qualities and by singular sensory representations of the same kind (modulo whether or not the singular contents of these representations are *successful*). Thus, assuming that the subjects have undergone the same perceptual learning, both experiences will introspectively seem to present a tomato, even though only the perception really does this.

Still, this commits us to saying that introspective appearances of presentation are sometimes misleading. So what, then, if the tomato-perception *introspectively seems* to present a tomato? Why think that this experience *really does* present a tomato? Here the rich pluralist will respond that introspective appearances can be *very* reliable without being *perfectly* reliable. Their account of the epistemology of presentation, given in chapter 8, is meant to fill out these ideas in a satisfying way.

There are two important virtues to the rich pluralist's response, which are best displayed against the backdrop of a competing naïve realist response. Many naïve realists hold that the tomato-perception and the tomato-hallucination generate similar introspective appearances, similar beliefs, and similar actions, but that they do so for extremely different reasons. By their natures, these experiences are radically different from one another.[1] This widespread naïve realist view has two important vices. First, it is a radical error theory about hallucinations: the account says that how they are is profoundly different from how they introspectively seem to be. Second, the account requires the naïve realist to provide very different explanations of a common set of introspective appearances, beliefs, and actions.

The rich pluralist replaces these vices with virtues. First, they honor *much*, although not *all*, of the appearance that the tomato-perception and the tomato-hallucination are the same by their very natures: they say that both experiences, by their natures, provide exactly the same kind of deep awareness of exactly the same sensory qualities and involve the same type of singular content (modulo the successfulness of that content). This is much better than a radical error theory.

Second, the tomato-perception and the tomato-hallucination cause similar introspective appearances, beliefs, and actions. The rich pluralist offers a simple, unified explanation of these facts: all of this occurs because both experiences involve identical kinds of deep awareness and identical kinds of singular representations (modulo their successfulness). This is a further virtue.

Thus, the rich pluralist has an especially effective response to the basic argument from hallucination.

11.2 A Screening-Off Argument

Naïve realism is famously subject to this *screening-off argument*:

B1. The tomato-perception and the tomato-hallucination will cause similar introspective appearances, beliefs, and actions.

B2. If the tomato-perception and the tomato-hallucination will cause similar introspective appearances, beliefs, and actions, then they belong to a common kind.

B3. If the tomato-perception and the tomato-hallucination belong to a common kind, then the acquaintance relation is explanatorily idle.

B4. If the acquaintance relation is explanatorily idle, then naïve realism is false.

B5. Naïve realism is false.[2]

What opens the naïve realist to this argument is that they identify a factor—acquaintance—that helps to explain certain introspective appearances, beliefs, and actions, while occurring only in the tomato-perception.

The rich pluralist embraces a similar commitment. They also identify a factor—this time it is *successful sensory representation*—that helps to explain certain introspective appearances, beliefs, and actions, while occurring only in the tomato-perception. Thus, the rich pluralist faces this parallel screening-off argument:

C1. The tomato-perception and the tomato-hallucination will cause similar introspective appearances, beliefs, and actions.

C2. If the tomato-perception and the tomato-hallucination will cause similar introspective appearances, beliefs, and actions, then they belong to a common kind.

C3. If the tomato-perception and the tomato-hallucination belong to a common kind, then successful sensory representations are explanatorily idle.

C4. If successful sensory representations are explanatorily idle, then rich pluralism is false.

C5. Rich pluralism is false.[3]

Let us examine the premises of the argument, starting with:

C1. The tomato-perception and the tomato-hallucination will cause similar introspective appearances, beliefs, and actions.

The idea is to compare the effects of these experiences in two subjects who are otherwise as similar as possible. The claim is not that the relevant introspective appearances, beliefs, and actions will be *identical*: the tomato-perception might result in a genuinely singular belief about an actual tomato, whereas the tomato-hallucination will result only in a failed singular belief. The claim is only that the effects will be *similar*. So understood, this premise is obviously true.

Next we have:

C2. If the tomato-perception and the tomato-hallucination will cause similar introspective appearances, beliefs, and actions, then they belong to a common kind.

The rich pluralist should grant this claim. For they accept the antecedent, which is simply C1. And they accept the consequent twice over. They think that any perception and any matching hallucination belong to *two* common kinds: both involve (i) deep awareness of exactly the same sensory qualities, together with (ii) singular sensory representations of the same kind, modulo whether their contents are successful. They even think that the antecedent and consequent are connected: they hold that the tomato-perception and the tomato-hallucination will have similar mental effects *because* these experiences belong to these common kinds.

We will get to C3 in a moment. First, however, skip ahead to:

C4. If successful sensory representations are explanatorily idle, then rich pluralism is false.

This is indisputable. It is just built into rich pluralism that successful sensory representations help to constitutively explain why typical perceptions present their targets and have strong referential character. Thus, the only move available to the rich pluralist is to deny:

C3. If the tomato-perception and the tomato-hallucination belong to a common kind, then successful sensory representations are explanatorily idle.

Fortunately, the rich pluralist has a clear justification for rejecting this premise. Yes, the tomato-perception and the tomato-hallucination cause *some* of the same introspective appearances, beliefs, and actions. The rich pluralist grants that the common kind explains these similarities. But there are also important differences between these experiences. The perception introspectively seems to present *that very tomato*, it positions the subject to form beliefs about *it*, and it positions the subject to act effectively on *it*. Nothing similar is true of the tomato-hallucination. These facts can be explained by appealing to something that is distinctive to the tomato-perception—say, by its being a *successful* sensory representation *of a particular tomato*.

Some have argued that such a response leads only to further trouble. The concern is that we want the factor that is distinctive to the perception—successful sensory representation, according to the rich pluralist—to help constitute the subject's *conscious perspective*. But if the tomato-perception and the tomato-hallucination belong to a common kind, then we cannot have what we want: presumably, it is this common kind that fully constitutes the subject's conscious perspective in both cases.[4] In other words:

C1*. If rich pluralism is true, then successful sensory representations do not in any way help to constitute the conscious perspective of the perceiving subject.

C2*. Successful sensory representations do in some way help to constitute the conscious perspective of the perceiving subject.

C3*. Rich pluralism is not true.

I observe that the expression "conscious perspective" is ambiguous. The rich pluralist's response to the argument depends on how this expression is understood.

Suppose first that we are discussing *hard* consciousness. Then C1* is clearly true. For the rich pluralist says that the subject's hard-conscious perspective is constituted purely by their deep awareness; successful sensory representations are irrelevant. However, chapter 4 contains extensive arguments against C2*: there I argued that deep awareness of the sensory qualities generates the hard problems of consciousness and the external world all by itself. Successful sensory representations *are* irrelevant to hard consciousness.

But suppose that we are discussing *strong access consciousness*. Then C1* is false. According to the rich pluralist, strong access consciousness is constituted by the particular kinds of deep awareness, sensory representations, and categorizations that the subject enjoys. In the case of the perceiving subject, this includes sensory representations that are distinctively *successful*. Thus, successful sensory representations *do* help to constitute their strong access-conscious perspective: they have strong access consciousness of a particular tomato *in virtue of* deploying sensory representations that successfully lock onto that very tomato.

Exactly the same response could be given if the argument were taken to concern weak access consciousness.

There is one final variant of the argument, which concerns *strong referential character*. C1* would then become:

C1**. If rich pluralism is true, then successful sensory representations do not in any way help to constitute the strong referential consciousness of the perceiving subject.

But this premise is clearly false. Strong referential consciousness is the kind of consciousness in virtue of which the subject can make *knowledgeable reference* to various targets. As explained in chapter 7, the rich pluralist holds that when a subject sees a red, round tomato, this kind of consciousness is constituted by two factors. One is successful sensory representation, which constitutively explains why the subject can make knowledgeable reference *to that very tomato*—it is because that tomato is what the subject's sensory representations lock onto. The second factor is deep awareness. This constitutively explains why the subject can refer to redness and roundness—it is because these are targets of deep awareness. The same factor also explains why the subject's reference to any of these targets can be *knowledgeable*—this occurs because deep awareness partly reveals the intrinsic characters of things, which explains the special sense in which the subject can know *what they are referring to.*

This blocks the final version of the screening-off argument.

Conclusion The rich pluralist's picture is this. The tomato-perception and the tomato-hallucination involve a shared kind of deep awareness and are thus identical with respect to hard consciousness. But when it comes to strong access consciousness, weak access consciousness, and strong referential consciousness, these experiences partly overlap and partly differ. The overlaps are explained by shared deep awareness, together with shared kinds of singular representations (modulo differences in whether those representations are successful). The differences are explained by *successful* sensory representation, which is unique to the tomato-perception.

On this picture, the common factor, deep awareness, lifts an entirely separate explanatory load from the uniquely perceptual factor, successful sensory representation. Thus, there is no danger that the common factor will screen off the uniquely perceptual one. When directed at rich pluralism, all versions of the screening-off argument fail.

11.3 An Argument from Introspective Access

There is one last argument from hallucination, the *argument from introspective access*, that threatens rich pluralism.[5] It begins by assuming the following claim for the sake of argument:

D1. The tomato-hallucination belongs to a positive, introspectible mental kind (call it the *common kind*).[6]

What is a *positive* mental kind? The basic idea is that membership in a positive mental kind is earned by how an entity *is* rather than by how it *fails to be*. As a foil, suppose that there is a mental kind defined by the property of *not* being introspectively discriminable from a veridical perception.[7] This would not be a positive mental kind. It is difficult to characterize the notion of a positive mental kind more precisely than this, since it seems possible in principle to characterize anything either in terms of how it is or how it is not. But, in a concessive spirit, I will assume that my opponent can solve this problem.

The next premise is:

D2. For any positive mental kind *K*, if the tomato-hallucination belongs to *K*, then the tomato-perception belongs to *K*.

D2 is plausible because membership in a positive mental kind is earned by how an entity *is*, not by how it is *not*. But everything that the tomato-hallucination positively is, the tomato-perception positively is: we have stipulated that these experiences are (partly or wholly) constituted by qualitatively identical neural events with qualitatively identical proximal causes. Perhaps the tomato-perception has additional positive aspects that are not found in the tomato-hallucination—perhaps only it presents a tomato. But the tomato-hallucination plausibly does not have positive aspects that are missing from the tomato-perception.[8]

D1 and D2 entail:

D3. The tomato-perception belongs to the common kind.

Now, surely the subject of the tomato-perception can introspectively tell that their experience belongs to the common kind. After all, the subject of the tomato-*hallucination* is supposed to be able to introspectively tell that *their* experience belongs to the common kind. Why would this ability vanish when the subject is veridically perceiving? Thus:

D4. If the subject of the tomato-perception can introspectively tell that their experience presents a tomato, then they have introspective access to two distinct aspects of their experience. (Supported by D3.)

In particular, they can introspectively tell that their experience belongs to the common kind and that it presents a tomato. These aspects must be distinct: the tomato-hallucination belongs to the common kind, but it does not present a tomato.

Now, it is immediately plausible that:

D5. If the subject of the tomato-perception has introspective access to two distinct aspects of their experience—one shared with hallucination, the other unique to perception—then they can selectively attend to each of these two aspects.

For when the subject introspects, surely, they can, in principle, direct their attention as they please. However:

D6. The subject of the tomato-perception cannot selectively attend to each of these two aspects.

This is meant to be immediately plausible: it seems that introspection does not turn up any extra aspect of perception to attend to. (If it did, we could introspectively distinguish the tomato-perception from the tomato-hallucination.)

Together, D4–D6 entail:

D7. The subject of the tomato-perception cannot introspectively tell that their experience presents a tomato.[9]

But all of this was premised on D1, the assumption that the tomato-hallucination belongs to a positive, introspectively accessible mental kind. Discharging this assumption, we reach our conclusion:

D8. If the tomato-hallucination belongs to a positive, introspectible mental kind, then the subject of the tomato-perception cannot introspectively tell that their experience presents a tomato.

If this argument is sound, then we must make a bitter choice. It strongly initially seems that the tomato-hallucination belongs to a positive, introspectible mental kind. It also strongly initially seems that the subject of the tomato-perception can introspectively tell that their experience presents a tomato. But, if D8 is true, then we must deny one of these appearances.

Many naïve realists simply accept this conclusion and make the bitter choice. In particular, they deny D1, the assumption that the tomato-hallucination belongs to a positive, introspectible mental kind. However, the rich pluralist embraces D1: they say that the tomato-hallucination belongs to a positive, introspectible kind constituted by deep awareness of sensory qualities such as redness and roundness, together with sensory representations of a particular object. I regard this as a great advantage of rich pluralism over the usual versions of naïve realism, for I find D1 to be extremely plausible on both theoretical and introspective grounds.

From a theoretical perspective, we want to explain why the tomato-hallucination will introspectively seem to present a mind-independent tomato, why it will introspectively seem to position the subject to make knowledgeable reference to a tomato, why it will position the subject to form a justified belief that "that thing" is a tomato, and why it will dispose a hungry subject to try to grab the apparent tomato. The simplest and most natural move is to appeal to a positive, introspectively accessible mental kind—preferably one that is shared by the tomato-perception, since the same phenomena will be found there. Other theoretical moves are possible. But they are less simple and less natural.

Then there is an introspective motivation for D1: the tomato-hallucination will introspectively seem to belong to a positive mental kind, *even to the subject who knows that they are hallucinating*. We must provide an error theory of *some* of the introspective appearances generated by this hallucination, and if necessary, we can do that here. But it would be better to honor the appearance.

So the rich pluralist accepts D1. How, then, can they respond to the argument?

They should grant D2–D5. In particular, they accept the following claims about the tomato-perception:

(i) it is hard-conscious (i.e., it is partly constituted by deep awareness),

(ii) it *purports* to make them sensorily aware of particulars (i.e., it is partly constituted by *putatively* singular sensory representations),

(iii) it *seems* to present particulars (i.e., it is categorized as presenting particulars),

(iv) it *really does* make them sensorily aware of particulars (i.e., it is partly constituted by *successful* singular sensory representations), and

(v) it *really does* present particulars (i.e., it is partly constituted by successful singular representations and deep awareness, woven together in a certain way).

They think that aspects (i)–(iii) are shared with the tomato-hallucination, while (iv)–(v) apply to the tomato-perception alone. They also think that an ordinary subject can introspectively tell that the tomato-perception has *all* of these aspects—not conceived via the theoretical descriptions in parentheses, of course, but via the ordinary, nonparenthetical descriptions.

The crucial premise, according to the rich pluralist, is:

D6. The subject of the tomato-perception cannot selectively attend to each of these two aspects (one shared with hallucination, the other unique to perception).

The rich pluralist rejects this claim: they think that the subject *can* selectively attend to each of the aspects mentioned above.

The rich pluralist acknowledges that some of these aspects are linked. For instance, aspect (iv), *actually* making the subject sensorily aware of particulars, requires aspect (ii), *purporting* to do so. Still, these aspects are plainly distinct. And even here, the rich pluralist says, the subject can attend to each aspect separately: there is a difference between noticing that an experience *purports* to make you aware of particulars and noticing that it *actually does this*.

This brings us to the crucial question. If the *perceiving* subject can introspectively single out an aspect of their experience, indeed two of them, that are distinctive to perception, then why can't the *hallucinating* subject tell that they are hallucinating by introspection alone? Why can't they simply notice that these two aspects are missing from their experience?

The rich pluralist's answer is that the hallucinating subject will *mistakenly think* that they can introspectively single out these two aspects that are unique to perception. For, first, their hallucination consists partly of *putatively but not actually* singular sensory representations. But such representations will of course introspectively seem to be *successful* singular sensory representations, just as a failed singular thought will introspectively strike its subject as a successful one. And, second, the hallucination will have a perceptual gestalt, so its subject will *mistakenly* categorize it as presenting a tomato. Thus, introspection alone will mislead them into thinking that they are perceiving rather than hallucinating.

To put things in another way, whether the subject is perceiving or hallucinating, they will have introspective access to their deep awareness and to their singular sensory representations (though not under those names). If they are perceiving, then these sensory representations will be *successful*, and their introspective access to them will position them to *know* that they are perceiving. If they are hallucinating, then these representations will be *unsuccessful*, but they will *mistake* them for successful ones. That will result in a *false belief* that they are perceiving.

There is one remaining puzzle. I have claimed that the subject of the tomato-perception can know, via introspection alone, that they are perceiving. But consider a subject who repeatedly flips between seeing the tomato in front of them and having a veridical hallucination of such a tomato. There will be no particular change in their beliefs when they transition from perceiving to hallucinating and back again—for example, they will be unable to identify the moments of transition. Does this not show that the subject of the tomato-perception *cannot* introspectively know that they are perceiving after all?

It does not. I concede that these unusual circumstances might act as defeaters, so that even if the subject of the tomato-perception *truly believes* that they are perceiving, *while these circumstances are at work*, they do not know that. But that is compatible with saying that *in normal circumstances*, the subject of the tomato-perception can know, via introspection alone, that they are perceiving. By analogy, in normal circumstances, I can know by vision alone that there is a tomato in front of me. But, if I am unknowingly flipping back and forth between really seeing a tomato and merely hallucinating one, then, even in those moments when I am truly seeing a tomato, my would-be knowledge that there is a tomato in front of me is defeated.

This removes the support for D6—again, the claim that the subject of the tomato-perception cannot selectively attend to each of these two aspects, one shared with hallucination, the other unique to perception. The argument is thereby blocked.

Conclusion There are good reasons, both theoretical and introspective, to think that the tomato-hallucination has at least one positive, introspectible aspect. It is to the detriment of many naïve realists that they reject it and to the credit of rich pluralists that they accept it.

Yet this opens the rich pluralist to the argument from introspective access: they must explain why the hallucinating subject cannot introspectively tell that they are hallucinating, simply by noticing that the distinctively perceptual aspects have gone missing. The rich pluralist's answer is that the hallucinating subject still deploys *unsuccessful* singular sensory representations, which will cause them to *mistakenly think* that the distinctive aspects are present. This creates the possibility of seamless transitions between the tomato-perception and the tomato-hallucination. *In these strange circumstances*, even the subject of the tomato-perception cannot know that they are perceiving. However, this does not undermine their ability to know this *in normal circumstances*.

Thus are the knots undone.

11.4 Conclusion

Rich pluralism has the resources to block the basic argument from hallucination, the screening-off argument, and the argument from introspective appearances. The theory accrues several advantages along the way. It can honor part of the appearance that the tomato-perception and the tomato-hallucination have the same essences. It can give a simple, unified explanation of the phenomenological, doxastic, and behavioral similarities between them. And it can affirm that there is a positive, introspectible mental kind that includes both experiences. This is all to the credit of rich pluralism.

Our long journey across this book is nearing its end. In the next chapter, I conclude things by retracing all of this book's main lines of argumentation.

12 Conclusion

Across the four parts of this book, I have offered complex arguments for rich pluralism. Part I used the framing assumption and extended partial revelation to construct an initial list of options, part II argued that rich pluralism was better than any of the representationalist options, part III argued that rich pluralism was better than any of the naïve realist (or strange pluralist) options,[1] and part IV solved some remaining problems for rich pluralism.

In this chapter, I look back on all of these arguments.

12.1 Part I Revisited: A List of Options

In chapter 1, when considering an ordinary case of seeing a mango, I introduced the following assumption:

> The *framing assumption*: When you see the mango, you are sensorily aware of it and its color and shape. In addition, whether you see or hallucinate a mango, you are not sensorily aware of any peculiar mental entities.

I went on to defend the following thesis:

> *Extended partial revelation*: In the first instance, all sensory experiences reveal a substantial portion of the essential truths about the sensory qualities and nothing else.

I argued that this thesis receives strong support from a direct examination of cases (chapters 2–3). I also argued that it is part of the best theory of which questions do, and do not, constitute hard problems of consciousness and the external world (chapter 4).

Together, the framing assumption and extended partial revelation greatly narrowed the field of viable theories.

One viable option was *impure revelatory monism*. This theory posits a single kind of sensory awareness, *sensing*, that targets external particulars like peppers, as well as sensory qualities like scarlet. According to the best version of this theory, sensing reveals the intrinsic characters of all its targets. It is then a priori that what is intrinsic to a sensory quality is essential to it, but this is not a priori for objects and events: this asymmetry is meant to explain extended partial revelation. Sensing can be understood as either representational or nonrepresentational; thus, impure revelatory monism comes in representationalist and naïve realist varieties.

There were also some viable pluralist options. The best pluralist theories posit two kinds of sensory awareness:

(i) Partly essence-revealing sensory awareness of the sensory qualities
(ii) Non-essence-revealing sensory awareness of particular objects, events, property-instances, and so on

Pluralistic representationalists take both kinds of awareness to be representational; *pluralistic naïve realists* take both to be nonrepresentational; *rich pluralists* treat (i) as nonrepresentational and (ii) as representational; *strange pluralists* occupy the remaining point in logical space.

A central conclusion of part I was that we should select one of these monistic or pluralistic options.

12.2 Part II Revisited: Rich Pluralism versus Representationalism

In part II, I compared rich pluralism to the remaining representationalist options, namely impure revelatory representationalism and pluralistic representationalism.

A central motivation for these theories is to account for hallucinations by saying that these experiences are in error about particulars. But the rich pluralist accepts the very same account of hallucinations, so the representationalist gets no edge here. Nor are the representationalist options more parsimonious than rich pluralism: even the impure revelatory representationalist must posit a second kind of awareness to account for hard-unconscious perceptions. Thus, these representationalists enjoy no major advantages over rich pluralists.

But they do suffer from a major disadvantage. They say that experiences consist partly of *representations* of the essences of the sensory qualities, which entails that experience can err on this topic. However, this is impossible: experience is infallible about the essences of the sensory qualities. The rich pluralist predicts this, for they say that our awareness of the sensory qualities is nonrepresentational.

I concluded that rich pluralism is superior to any representationalist alternative.

12.3 Part III Revisited: Rich Pluralism versus Naïve Realism

Part III compared rich pluralism to the remaining naïve realist options—impure revelatory naïve realism and pluralistic naïve realism.

Naïve realism is held up by four main pillars. It has been thought that this theory has the best account of why hard-conscious perception (i) reveals the *intrinsic character* of its targets, (ii) positions us to make *knowledgeable reference* to its targets, (iii) *presents* its targets, and (iv) *introspectively seems* to present its targets.

When considering these ideas, I accepted part of the naïve realist orthodoxy: I agreed that hard-conscious perceptions position us to make knowledgeable reference, and present their targets, *because* they reveal the intrinsic characters of external particulars. However, I argued that the data extend beyond this. Not only hard-conscious perceptions, but *all* experiences, also reveal the intrinsic characters of certain universals—the sensory qualities. I showed that this core explanatory idea is not just compatible with naïve realism. It is also compatible with representationalism and rich pluralism. All that we must do is treat experiential awareness of sensory qualities as primitive, rather than providing a physicalist reduction. That removes the first three pillars of support for naïve realism.

Then there is the fourth pillar: why do hard-conscious perceptions (and hallucinations) introspectively seem to present their targets? I argued that this occurs because we *categorize* our experiences as, for example, presenting mind-independent targets based on an overall gestalt. This account, the *categorization account*, is again available to all comers.

Thus, the four pillars do not support naïve realism in particular. They equally support rich pluralism (and, for that matter, impure revelatory representationalism and pluralistic representationalism). This leaves naïve realists with no significant advantages over rich pluralists.

But rich pluralists have some major advantages over naïve realists.

Only four versions of naïve realism provide adequate accounts of hallucination. The first is *flexible naïve realism*, which says that hallucinations involve awareness of mind-dependent instances of the sensory qualities. However, this theory posits instances of the sensory qualities that do not inhere in anything, not even a region of space, and it accepts metaphysical conspiracies—it adds new commitments to cancel any distinctive predictions that it would otherwise make. Rich pluralism is free of these costs.

The second naïve realist theory of hallucination is the *imaginative view*, which says that hallucinations are a kind of involuntary sensory imagining. However, this theory ends up positing three very different kinds of mental events—acquaintance, covert awareness, and sensory representation—to account for hard-unconscious perceptions, hard-conscious and hard-unconscious hallucinations, and sensory imaginings. The rich pluralist posits only deep awareness and sensory representations. In addition, the imaginative theorist says that it can be impossible in principle to discriminate the hard characters of perceptions and hallucinations, *even though* these hard characters are of radically different kinds—the former is constituted by worldly objects and property-instances, the latter by something entirely different. Rich pluralists do not say that radically different hard characters can be indiscriminable in this way. Finally, the imaginative theorist must say these two in-principle indiscriminable hard characters just so happen to be underpinned by a common neural event, *despite* their radical differences in kind. The rich pluralist can accept the simpler view that the common neural event underpins a single kind of hard character.

Then there is the *quality-awareness view*, which says that what it is to hallucinate is to sense the sensory qualities alone, without sensing any particulars. If we accept this view, we must posit at least three basic kinds of mental events: *acquaintance*, which constitutes hard-conscious perceptions (and relates us to the sensory qualities in hard-conscious hallucinations); *covert awareness*, which constitutes hard-unconscious perceptions; and *erring*, which constitutes our taking particulars to exist in hard-conscious and hard-unconscious hallucinations. Rich pluralism is more parsimonious.

The fourth and final naïve realist theory is *pluralistic naïve realism*. Like rich pluralists, these theorists posit two kinds of awareness constitutive of hard-conscious perception; unlike rich pluralists, they say that both kinds are nonrepresentational. Pluralistic naïve realists are forced to posit no fewer than four different kinds of mental events. They already posit *acquaintance* with particulars and *deep awareness* of sensory qualities. They must also posit *erring*, to explain why hallucinations seem to make us aware of particulars, and *sensory representations*, to account for hard-unconscious imaginings. Rich pluralism shoulders the same explanatory work with just two posits.

My conclusion was that rich pluralism is superior to any version of naïve realism.

12.4 Part IV Revisited: Remaining Problems

In part IV of the book, I engaged with two clusters of problems. First, I considered a deep dilemma about phenomenal character, consisting of this pair of plausible but inconsistent claims:

PC1. Any perception and matching hallucination will have exactly the same phenomenal character.

PC2. It is not the case that any perception and matching hallucination will have exactly the same phenomenal character.

I argued that the expression *phenomenal character* has been used with different concepts in mind. Some use this expression with *hard character* in mind, in which case the rich pluralist will say that PC2 is false: hard character is constituted by deep awareness of sensory qualities, which will be the same across any perception and matching hallucination. Others use the expression *phenomenal character* with *strong referential character* or *strong accessible character* in mind. If we adopt this understanding, then PC1 is clearly false: seeing a scarlet pepper positions me to make knowledgeable reference to it, and this mental event is a hard-conscious one that makes me sensorily aware of the pepper. Neither of these things can be said of a matching hallucination.

These solutions are extremely attractive: each one lets us reject a premise that, once properly understood, is not even prima facie plausible. That is a great advantage.

I then offered my replies to a series of arguments from hallucination against rich pluralism. One key defensive idea was that the rich pluralist honors both the *similarities* and the *differences* between hard-conscious perceptions and matching hallucinations. Regarding the similarities, the rich pluralist says that both mental events involve deep awareness of the same sensory qualities and (putatively) singular sensory representations of the same kind. That is why, in both cases, the subject will *introspectively seem* to be presented with a pepper and its instances of scarlet. Regarding the differences, the rich pluralist says that only the perception involves a *successful* singular representation of an actual pepper and its instances of scarlet.

Thus, only it genuinely *presents* the subject with these things, and only it makes them *strongly access-conscious* of them.

Thus, I did not find any further objections to rich pluralism.

12.5 Conclusion

We can view the basic arguments of this book from an even higher altitude. The idea is that an adequate theory of perception must accept that experience is partly revelatory, in a particular sense (part I). Rich pluralism is better than any adequate representationalist theories, because the latter mistakenly predict that experience is fallible about the very essences of the sensory qualities (part II). And rich pluralism has the same explanatory power as any adequate naïve realist theories, but rich pluralism is, among other things, more parsimonious (part III). What is more, rich pluralism can dissolve the deep dilemma about phenomenal character and block a series of arguments from hallucination (part IV).

It is worth ending with a synoptic observation. Intuitively, there are many unities tying together experiences of various kinds. Rich pluralism honors them all. In particular:

There seems to be something that unites all sensory mental events, whether hard-conscious or not: the pluralist explains this in terms of sensory representation. There seems to be something that unites perceptions and genuine episodic memories—both make the subject sensorily aware of particulars. The rich pluralist explains this in terms of *successful* sensory representations. There seems to be something that unites all hard-conscious events: the rich pluralist explains this in terms of deep awareness. There seems to be something that unites all mental events with apparent presentational character: the rich pluralist explains this in terms of categorization via an overall gestalt. Yet there also seems to be something, namely *genuine* presentation, that is distinctive to hard-conscious perceptions: the rich pluralist says that this is a matter of deploying a sensory representation that represents via a perceptual relation and that is lit up with deep awareness.

This package of views is explanatorily powerful. Yet, fundamentally speaking, it requires only two distinctive posits: deep awareness and sensory representation.[2] Thus, its name notwithstanding, rich pluralism is surprisingly parsimonious.

I conclude that it is the best theory of perception to date.[3]

Glossary

Accidental truth (about *x*). For a truth *T* about some entity *x* to be *accidental* is for *T* not to be an essential truth about *x*. An accidental truth about *x* might be an essential truth about some other entity *y*, and accidental truths may still be necessarily true.

Acquaintance reveals intrinsic character (ARIC). This is the thesis that every acquaintance relation constitutive of hard-conscious perception reveals a substantial portion of the intrinsic character of its targets.

Apparent presentational character. For a mental item to have *apparent presentational character* is for it to introspectively seem to present one or more particulars. See "presentation" and "the categorization account."

Apparent presentationality. This thesis consists of the following claims. First, any typical hard-conscious perceptual experience or sensation has *apparent presentational character*—that is, it introspectively seems to present one or more particulars. In perceptual experience, these particulars typically seem to exist independently of the subject's mind. In sensation, these particulars typically seem to exist in a way that depends on the subject's mind. Second, apparent presentational character is not found in typical hard-unconscious perceptions, typical hard-conscious episodic memories or sensory imaginings, or typical singular thoughts.

Categorization account (of apparent presentational character). The account is this. For an experience to introspectively seem to present a target, and for the target to introspectively seem to be mind independent or mind dependent, is for the subject to *categorize* the experience or its targets as such. In normal, mature human beings, this categorization occurs on the basis of an overall gestalt.

See also "gestalt."

Causal content account (of apparent presentational character). According to this account, for a mental item to have apparent presentational character toward a target is for it to represent its target as standing in a certain causal relation to that very mental item.

Cognitive view. This is a naïve realist view of hallucination. It says that to hallucinate something is to undergo a mental event that lacks phenomenal character but that produces the same cognitive effects that a veridical perception of that thing would have produced in a rational subject with the same background mental situation.

Collective full revelation. This is the thesis that consciously perceiving any number of sensory qualities reveals all *collective* essential truths about them and does not mistakenly seem to reveal any *collective* essential falsehoods about them.

Conservatism about hard character. This is the view that, of the properties and relations that we perceive, only low-level properties and relations—color, shape, texture, and so on—constitute the hard character of our perceptual experiences. For a contrary view, see "liberalism about hard character."

Content. See "representation."

Content constitution account. This account says that for a mental item to have strong referential character toward some target *T* is for it to have a content that contains *T* as a constituent.

Context sensitivity, shallow and deep. For a context-sensitive expression to be *shallow* is for it to have a reasonably narrow range of possible referents. If a context-sensitive expression is not shallow, then it is *deep*.

Deep awareness. See "rich pluralism."

Deep context-sensitivity view. According to this view, any *F*-entailing claim of the form "what *x* is like is *F*" is deeply context sensitive, because it carries the meaning that *F* is a property of *x* that is relevant in the context at hand (where this context may be further specified with prepositional phrases and subordinate clauses). See also "context sensitivity, shallow and deep."

Determinate content account (of apparent presentational character). This account says that what it is for a mental event to introspectively seem to present a mind-independent particular is for it to be a sensory representation with a determinate content.

E-theory. This is a theory of what generates the hard problems of consciousness and the external world. It consists of the following claims:

> *Extended partial revelation*: In the first instance, all sensory experiences reveal substantial essential truths about the sensory qualities and nothing else.

E1. Often, it strongly initially seems impossible to us for purely physical entities to satisfy the essential conditions that are revealed to us in sensory experience (as described in extended partial revelation). When and because this happens, there are hard problems.

E2. It strongly initially seems impossible to us for purely physical facts to make it the case that there exists partly revelatory awareness (as described in extended partial revelation). When and because this happens, there are hard problems of consciousness.

E3. There are no other sources of hard problems of consciousness or the external world.

Easy account (of apparent presentational character). The account is this. Take any subject whose experience presents them with a mind-independent/mind-dependent target and who possesses the concepts of presentation and mind independence/mind dependence. Just on the basis of careful, rational reflection on their experience, it will introspectively seem to them that they are presented with a mind-independent/mind-dependent target.

Essence. The essence of any entity *x* is what it is to be *x*.

Essential truth. An *essential truth* about some entity *x* is any truth of the form, <It lies in the essence of *x* that . . .>.

Existential content. For a content to be *existential* is for it not even to be directed at a particular. For a contrast, see "singular content."

Extended partial revelation. This is the thesis that, in the first instance, all sensory experiences reveal a substantial portion of the essential truths about the sensory qualities and nothing else.

Flexible naïve realism. This is a version of naïve realism that accepts the following commitments. First, that sensory qualities like redness are *ontologically flexible*: the existence of an instance of redness can be fully grounded *either* in the existence of an ordinary object such as a tomato *or* in a subject's being in a certain internal state (perhaps a brain state). Second, that to consciously hallucinate a red tomato is, in part, to be acquainted with a perfectly real but *mind-dependent* instance of redness. Third, that the existence of this instance of redness is fully grounded in the fact that the subject is in a certain internal state, and the instance does not inhere in anything—not even the subject's mind or experience. Fourth, that to consciously perceive a red tomato is to be acquainted with an instance of redness whose existence is *ontologically overdetermined*: its existence is fully grounded in the existence of the subject's internal state *and, separately*, in the existence of the tomato.

Framing assumption. This is the assumption that, when you hard-consciously see the mango, you are sensorily aware of it and its color and shape; in addition, whether you hard-consciously see or hallucinate a mango, you are not sensorily aware of any peculiar mental entities.

Full revelation. This thesis is the conjunction of the following two claims:

Self-intimation: Consciously perceiving any sensory quality Q reveals every truth of the form <It lies in the essence of Q that . . .>.

Infallibility: Consciously perceiving any sensory quality Q does not mistakenly seem to reveal any false claims of the form <It lies in the essence of Q that . . .>.

Gestalt. A *gestalt* is a weighted set of cues (e.g., cues involving the sensory qualities). A *perceptual gestalt* is a gestalt taken to indicate that the subject is presented with a mind-independent target. A *sensational gestalt* is a gestalt taken to indicate that the subject is presented with a mind-dependent target. A *presentational gestalt* is a gestalt taken to indicate that the subject is presented with *something*, whether mind independent or mind dependent.

Grounding (partial and full). For the fact that *p* to *partly/fully ground* the fact that *q* is for *q* to hold partly/fully in virtue of *p*, metaphysically speaking.

Hallucination. According to the rich pluralist, what it is to be a *hallucination* is to be a sensory representation that is *mistakenly* taken to be a perception. In normal, mature human beings, this occurs when and because the sensory representation, despite not being a perception, is taken to have a perceptual gestalt.

Hard character. This property is defined in relation to the following problems:

1. The *explanatory gap problem*: We seem unable to understand how the experience could be nothing more than something that is conceived purely physically and/or functionally.
2. The *knowledge problem*: We seem unable to know everything about the experience just on the basis of having all physically and functionally conceived information.
3. The *conceivable absence problem*: We seem to be able to conceive of the experience as present in one subject *S* yet absent in some other subject who is identical to *S* in all functional and physical respects.
4. The *conceivable inversion problem*: We seem to be able to conceive of the experience as invertible across subjects who are identical in all functional and physical respects.

In particular, if there is a single natural kind property of experience that tends to generate enough of problems 1–4 in paradigmatic cases, then the *hard character* of a mental item is the item's property that belongs to this natural kind. If there is no natural kind that generates enough of problems 1–4, then *hard character* simply refers to any property of that experience that generates enough of these problems. If there several natural kinds that generate enough of problems 1–4, then *hard character* refers indeterminately to these.

According to the rich pluralist, for a mental item to have hard character is for it to be at least partly constituted by deep awareness (i.e., partly essence-revealing awareness of sensory qualities).

Hard-character externalism. This is the view that the hard character of some experiences does not supervene on the intrinsic properties of the subject.

Hard-character internalism. This is the view that the hard character of every experience supervenes on the intrinsic properties of the subject.

Hard consciousness. For a mental item to be *hard-conscious* is for it to have hard character.

Hard problem. For there to be a *hard problem* is for the following to be true:

(i) we have strong background evidence that *p* fully grounds *q*—but, at the same time,
(ii) it strongly initially seems impossible to us that *p* fully grounds *q*: we cannot see why *p* would make it the case that *q* rather than making it the case that *r*.

Holistic revelation. This is the thesis that consciously perceiving all of the sensory qualities within a given quality space reveals all essential truths about those sensory qualities and does not mistakenly seem to reveal any essential falsehoods about them.

Illusion. According to the rich pluralist, for the subject to have an *illusion* is for them to successfully sensorily represent some particular, *o*, and deploy a singular representation of some property-instance *F* of *o*, but to misrepresent *F* as being an instance of some sensory quality Q when in fact it is not. This illusion is *hard-conscious* if, in concert with this, the subject is deeply aware of Q.

Imaginative view. This is a naïve realist view of hallucination. It says that part of what it is to be a hallucination is to be an involuntary sensory imagining.

Impure revelatory monism. According to this view, to consciously perceive is to deploy exactly one kind of awareness, *sensing*, that targets particulars and the sensory qualities, understood as universals. However, in the first instance, sensing only reveals a substantial portion of the essences of the sensory qualities, not of anything else.

Impure revelatory naïve realism is the version of this view that treats sensing as nonrepresentational. *Impure revelatory representationalism* is the version that treats sensing as representational.

Impure revelatory naïve realism. See "impure revelatory monism."

Impure revelatory representationalism. See "impure revelatory monism."

Indiscriminability view. This is a naïve realist view of hallucination. It says that to hallucinate is to be in a mental situation that (i) is not a perception of any kind but (ii) cannot be introspectively discriminated from a veridical perception.

Individual full revelation. This is the thesis that consciously perceiving any sensory quality reveals all *individual* essential truths about it and does not mistakenly seem to reveal any *individual* essential falsehoods about it.

Infallibility. See "full revelation."

Intrinsicality datum. This is the datum that sensory experiences of any kind—including hard-conscious perceptions, hallucinations, sensations, sensory imaginings, and episodic memories—reveal that it lies in the essences of certain *properties* that they are intrinsic and that anything with those properties is intrinsically a certain way. In addition, hard-conscious *perceptions* reveal that perceived *objects* are intrinsically a certain way but not whether they are essentially that way.

Involuntary determinate content account (of apparent presentational character). This account says that what it is for a mental event to introspectively seem to present a particular, whether mind independent or mind dependent, is for it to be an involuntary sensory representation with a determinate content.

Liberalism about hard character. This is the view that the hard character of some of our perceptual experiences is constituted partly by *high-level properties and relations*: for instance, by causation, natural kind properties, emotional properties, or affordances. For a contrary view, see "conservatism about hard character."

Mental object view. This is the view that to hallucinate is to be acquainted with a mental object.

Modified determinate content account. This account says that what it is for a mental event to introspectively seem to present a particular, *whether mind independent or mind dependent*, is for it to be a sensory representation with a determinate content.

Monism. View that what it is for a subject to consciously perceive is for them to deploy exactly one kind of sensory awareness.

Naïve realism. View that (i) to perceive is to deploy at least one kind of awareness, typically of external particulars and their properties/property-instances, and (ii) every kind of awareness constitutive of perception is nonrepresentational (i.e., immune to a particular kind of error), as described in "representation."

Nonsensory states/events. See "sensory events."

No revelation. This is the thesis that consciously perceiving any sensory quality Q does not reveal any essential truths about Q, that is, truths of the form <It lies in the essence of Q that . . .>.

Partial revelation. This is the thesis that consciously perceiving any sensory quality Q reveals a substantial portion of the essential truths about Q, that is, truths of the form <It lies in the essence of Q that . . .>.

Perception. According to the rich pluralist, what it is to be a perception is to be a sensory representation that represents by standing in a perceptual relation to one or more targets.

A *perceptual relation* is a relation between a subject's sensory representations and the target(s) such that (i) the representations represent the target(s) directly, (ii) the representations update in real time, and (iii) the contents of the representations vary with the target's spatial position relative to some part of the subject.

The *perspectival content account* (of apparent presentational character). The account is this:

(i) Almost all typical visual perceptions represent the target as independent of the subject and as perspectivally connected to the experience. Almost all typical visual sensations do not do this.

(ii) This difference in content constitutes the obvious phenomenological difference between typical visual perceptions and typical visual sensations.

Pluralism. This is the theory that what it is for a subject to consciously see, for example, an ovoid orange mango is for them to deploy two very different kinds of sensory awareness in concert. I assume that pluralists will posit the following two kinds of awareness in particular:

(i) Partly essence-revealing awareness of the sensory qualities
(ii) Non-essence-revealing awareness of particulars

Pluralism comes in four major varieties. *Pluralistic representationalists* treat (i) and (ii) as representational. *Pluralistic naïve realists* treat (i) and (ii) as nonrepresentational. *Strange pluralists* treat (i) as representational and (ii) as nonrepresentational. *Rich pluralists* treat (i) as nonrepresentational and (ii) as representational, but they accept many further commitments, as well—see "rich pluralism."

Pluralistic naïve realism. See "pluralism."

Pluralistic representationalism. See "pluralism."

Presentation. This is a relation that introspectively seems to occur in typical hard-conscious perceptions, hallucinations, and sensations and that introspectively seems *not* to occur in thoughts, sensory imaginings, or episodic memories. See also "apparent presentational character"; "apparent presentationality."

Quality-awareness view. This is a naïve realist theory. It says that what it is to perceive is to be acquainted with particulars instantiating sensory qualities, understood as universals; what it is to hallucinate is just to be acquainted with sensory qualities, without being acquainted with any particulars.

Representation. A representation is an entity that is *about* something—a *content.* The notion of a content is taken as primitive for present purposes. I do not assume that all contents are metaphysically possible.

I assume that any natural, broad representational kind (belief, desire, sensory experience) contains some representations with contents that obtain and some representations with contents that do not obtain—even if some *particular* contents obtain or fail to obtain necessarily. In this sense, representations are entities of a kind that can go right or wrong.

See also "existential content"; "singular content."

Representationalism. This is the view that (i) to perceive is to deploy at least one kind of awareness, typically of external particulars and their properties/property-instances, and (ii) every kind of awareness constitutive of perception is representational (i.e., subject to a particular kind of error), as described in "representation."

Revealing. For a mental state or event *M* to *reveal* a truth *p* about some entity *e* is for the following conditions to be met:

(i) *M* makes the subject aware of entity *e*.
(ii) For a conceptually sophisticated subject, rational reflection on *M* will, by itself, make it seem that *p* is true.
(iii) The subject's awareness of entity *e*, as described in (i), provides an intelligible basis for its seeming that *p* is true, as described in (ii).

Revelation. See "full revelation." For a related but weaker thesis, see "partial revelation."

Revelation of necessities. This is the thesis that all sensory experiences reveal truths of the form <Necessarily, sensory quality *Q* is *F*>.

Rich pluralism. The very specific pluralistic theory developed in this book. Some of its main commitments are these.

Regarding *deep awareness*:

DA1. Deep awareness occurs in all *sensory experiences*, including all conscious perceptions, hallucinations, sensory imaginings, genuine episodic memories, and merely apparent episodic memories.
DA2. At least in the sensory experiences of beings like us, the only targets of deep awareness are certain universals—the *sensory qualities*.
DA3. Deep awareness is a nonrepresentational form of awareness-of that reveals to the subject a substantial portion—not necessarily all!—of the essences of its targets.

Regarding *successful sensory representation*:

SSR1. Successful sensory representation occurs in all perceptions and in all (genuine) episodic memories. It does not occur in any hallucinations, sensory imaginings, or failed episodic "memories"—here we find only failed sensory representations.
SSR2. The targets of successful sensory representation include particulars, such as ordinary objects and their property-instances. (Rich pluralism is neutral on whether these targets sometimes include universals, too.)
SSR3. Successful sensory representation is a representational form of awareness-of that never even seems to reveal any portion of the essences of its targets.

Self-intimation. See "full revelation."

Sensing. See "impure revelatory monism."

Sensory events. These include hard-conscious and hard-unconscious sensory perceptions, hallucinations, sensations, episodic memories, and sensory imaginings. They stand in contrast to *nonsensory* states or events, such as thoughts, intentions, and desires.

Sensory representations. These are posited by some theorists, including representationalists and rich pluralists, to account for sensory events. See "rich pluralism"; "sensory events."

Simple full revelation. This is the thesis that consciously perceiving any sensory quality reveals all essential truths about it and does not mistakenly seem to reveal any essential falsehoods about it.

Singular content (successful and failed). For the content of a representation to be *singular* is for the representation to be putatively directed at a particular, such as an object, property-instance, or event. The expression *singular content* refers even to contents that are only putatively singular.

Strange pluralism. See "pluralism."

Strong accessible character. If a mental item is hard-conscious, then its *strong accessible character* is its maximal property of the form, *making . . . directly available to its subject.* If a mental item is not hard-conscious, then it has no strong accessible character. Thus, what it is for a mental item to have strong accessible character—equivalently, to be *strongly access-conscious*—is for it to be hard conscious and to make something directly available to its subject.

Strong referential character. For a mental item to have *strong referential character* toward something is for it to position the subject to make *de re* reference to that thing *and* to position them to know, in a certain special sense, what they are thereby referring to.

Relatedly, for a mental item to have *super-weak referential character* toward an entity is for it to position the subject to refer to that entity. For a mental item to have *weak referential character* toward an entity is for it to position the subject to make *de re* reference to that entity.

Successful singular contents actually include one or more particulars—for example, the content <Socrates is a philosopher>, which includes Socrates himself. A representation that is directed at a particular without actually latching onto one has a *failed* singular content—for example, the content of a failed demonstrative term.

For a contrast, see "existential content."

Successful sensory representation. See "representation"; "rich pluralism."

Super-weak referential character. See "strong referential character."

Transparency. This is the following thesis. Take any typical perceptual experience and consider the properties of which the subject is directly aware. All of these properties are experienced as properties *of ordinary particulars* that seem to exist independently of the subject's mind; none of them are experienced as properties *of the experience itself.*

Weak accessible character. By definition, the *weak accessible character* of a mental item is its maximal property of the form, *making . . . directly available to its subject.* Thus, what it is for a mental item to have weak accessible character—equivalently, to be *weakly access-conscious*—is for it to make *something* directly available to its subject.

According to the pluralist, the *weak accessible character* of an experience is the property of making all of the following directly available to the subject:

(i) the targets of sensory representations (if these representations are weakly access-conscious),
(ii) the sensory qualities that are the targets of deep awareness (if the experience is partly constituted by weakly access-conscious deep awareness), and
(iii) the contents of any categorizations of these targets (if this categorization occurs within experience proper).

Weak referential character. See "strong referential character."

Weakened easy account (of apparent presentational character). The account is this. Take any subject whose experience presents them with something and who possesses the concept of presentation. Just on the basis of careful, rational reflection on their experience, it will introspectively seem to them that they are presented with something.

Notes

Chapter 1

1. Some would question the reliability of introspection: see, for example, Schwitzgebel (2008, 2012). I grant that introspective verdicts vary in their reliability, but the introspective judgments that are central to this book—especially those concerning revelation—strike me as very hard to dispute. On this point I agree with Goff (2017, chap. 5, esp. pp. 111–112).

1. Although this theory has mostly fallen out of favor, a recent defense appears in D'Ambrosio (2019).

2. It also rules out the theory of Nanay (2012a), which says that conscious perception makes us aware of tropes but not objects.

3. This theory has its roots in the work of Thomas Reid. More recently, it has been defended in Quilty-Dunn (2015).

4. To make just one suggestion, we might stipulate that for an instance of awareness to be *direct* is for it not to be epistemically based on any other instance of awareness. To lock onto the notion of *epistemic basing*, consider a subject who (i) knows that *p*, (ii) knows that if *p*, then *q*, and (iii) arrives at the conclusion that *q*. But suppose that they do not conclude that *q* by reasoning from their beliefs that *p* and that if *p*, then *q*. They just draw the conclusion that *q* from an irrational hunch that *q*. *Epistemic basing* is the relation that obtains between this subject's irrational hunch and their belief that *q*; it is also the relation that fails to obtain between their belief that *q* and their beliefs that *p* and that if *p*, then *q*.

By understanding direct awareness in terms of epistemic basing, we can make the distinction that we want. Suppose again that I see a shadow, but not the person casting it. I am directly aware of the shadow but only indirectly aware of the person: my awareness of the person is *epistemically based on* my awareness of their shadow.

5. Mental paint theories are defended in Block (1996, 19–20), Millar (2017, 134–135), and Beck (2019, sections 3, 6, 8).

6. This is part of the *transparency thesis* defended in Harman (1990) and Tye (2002). For reasons explained in section 7.5 in chapter 7, I am not committed to the full-strength version of the transparency thesis.

7. It might seem that Tye (2009) develops a pluralistic version of representationalism. For he thinks that conscious perception makes us sensorily aware of ordinary objects and, independently, of certain universals. These two kinds of sensory awareness come apart sometimes—for instance, in hallucination, we are only aware of universals.

Although Tye comes close to pluralism, I do not think he quite makes it there. The problem is that although he posits *independent* sensory awareness of ordinary objects and of universals, he does not make any significant distinction in the *kind* of awareness across the two cases. In particular (all references below are from Tye 2009):

- Tye treats both kinds of awareness as representational: he suggests that objects and properties both figure in the content of veridical perceptions (82).
- Tye treats both kinds of awareness as being constituted by nonconceptual content: "The picture of visual experience I wish to recommend, then, is one under which the content of visual experience is robustly nonconceptual" (108).
- Tye treats both kinds of awareness as affording acquaintance with their targets: "My notion of acquaintance can be illustrated by example. I am acquainted with the color red, the city of Athens, the Apple computer at which I am now typing, . . . I have encountered (or am now encountering) all these things in my experience" (101).
- Tye argues on pp. 188–189 that because we are acquainted with colors, we are immune to a certain kind of error about them. But, as noted in the previous point, Tye also thinks that we are acquainted with objects. Thus, he is committed to thinking that we are immune to the same kind of error about them.

There remains one apparent asymmetry. Tye says that universals constitute phenomenal character, when we are aware of them, but particulars do not, even when we are aware of them. However, this distinction does not seem to play any particular role on Tye's theory: I cannot find any important theoretical difference, within Tye's theory, between phenomenal characters (consisting of universals, when we are aware of them) and objects (when we are aware of them).

Thus, as far as I can tell, Tye posits two instances of a single kind of awareness. The result is a monistic theory, not a pluralist one.

8. As it happens, rich pluralism contains commitments not only about perception but also about hallucination, sensory imagination, episodic memory, and more.

Chapter 2

1. In fact, I suspect that the sensory qualities are not ordinary qualities like scarlet but less familiar *appearance properties* that bear some interesting relation to ordinary qualities. It is easy enough to modify my claims throughout this book to accommodate this point. Still, for the sake of simplicity, I will glide over this complication.

2. Later in the book, I will distinguish several kinds of consciousness. What I have in mind here is what I will eventually call *hard consciousness*.

3. For some descriptions of this phenomenological point, see Levine (2001, 84), Levin (2007, 96–97), Johnston (1992, 223–225), and Schroer (2010, 510). Some of these theorists put the datum not in terms of sensory qualities like red but in terms of "phenomenal properties" and the like.

4. My entire discussion of what essences are has stuck pretty closely to Fine (1994). For Fine's comparison of essential and metaphysically necessary truths to definitional and analytic truths, see pp. 10–14.

5. The literature on essences is vast. For just a few examples, see Fine (1994), Leslie (2011), Denby (2014), Correia and Skiles (2019), and Spencer (2020).

6. I hold that revelation, at least when it occurs in experience, occurs because of *sensory* awareness. You might worry about this: since we might enjoy the relevant seemings only after performing complex feats of reasoning, doesn't revelation really occur because of *rational* awareness?

It does not: a mental state or event reveals a truth as long as it provides an *intelligible basis* for extracting certain truths—as long as it *positions* the subject to extract those truths. The subject need not be able to *take advantage* of that position. Thus, even subjects without concepts—subjects who surely lack rational awareness of anything—can have truths revealed to them.

Still, isn't it perverse to characterize *sensory* awareness in terms of *rational* reflection? No: it is often useful to characterize the epistemic power of a mental state or event in terms of its consequences given rational reflection. For instance, suppose that someone gives me some testimony *p*. For certain purposes, it is useful to say that what I have learned is not just *p*, but anything that follows from *p* together with my prior information.

7. Some may wish to add that the subject must first possess the concept of essences. I myself find this proviso unnecessary: it seems to me that the concept of essences can be acquired a priori, without the need for any experiences.

8. Scientists have made great progress on characterizing the dimensional spaces associated with colors and sounds. They made much less progress doing the same for smells. Thus, you might object that smelling vanilla does not really reveal any essential truths about that scent.

To this I have three responses. First, smell is by far our weakest sense, so it would be unsurprising if we had more difficulty characterizing the dimensional space associated with it. Second, I still find it immediately plausible that smelling vanilla reveals a substantial portion of the essence of that scent. Just think about how much would be lost if you could not smell vanilla, even if you had a detector that listed for you all nearby scents. Finally, normal perceivers can readily tell that the scent of vanilla is more similar to the scent of oak than to the scent of black pepper.

9. I see sympathy for no revelation in McLaughlin (2003), Campbell (2005), Byrne and Hilbert (2007), Tye (2009), Allen (2011), and Broi (2020).

10. For those who would resist this last claim, I could simply run all of the arguments of this book in terms of color-like properties rather than colors.

11. As Ivanov (2019, section 2) observes.

12. Contemporary interest in full revelation is due to Mark Johnston (1992, 223–228). The thesis is usually restricted to vision, but I find the restriction unnecessary: the idea is equally plausible when applied to the experiences of hearing a violin or smelling vanilla. Thus, my formulation will speak of sensory qualities in general. Full revelation is also defended in Ivanov (2019) and Kappes (2020). David Lewis (1995, 141–142) discusses a related thesis, except that the allegedly revealed truths are about experiences rather than sensory qualities. I take no stance on Lewis's thesis.

13. This objection is carefully developed in Broi (2020), although his focus is on *phenomenal properties*, which he stipulates to be properties of experiences, rather than on sensory qualities.

14. Broi (2020, 30–33).

15. Broi (2020, 33–35).

16. This is the central proposal of Kappes (2020).

17. Kappes (2020, 806) mentions this thesis while remaining neutral about its truth.

18. It is difficult to characterize the type of composition at issue here. A salad might be composed of many different vegetables, but the elements of the salad are separable in a way that does not seem to be true of the elements (say, red and yellow) of a compound color (orange).

19. These points are laid out in detail in Allen (2011, 163–167), although his focus is on something like simple full revelation. Note that the objection works just as well regardless of which colors are elemental and which are compound. For, in each of the cases discussed, the introspective seemings are split.

20. It is worth mentioning that the objection from the elemental-compound distinction is an instance of a more general objection that is developed in McLaughlin (2003, 98): the objection that there are surely *some* essential truths about color that

can only be learned through detailed scientific inquiry (where just getting a good look at all of the colors does not count). Holistic revelation and all of the versions of full revelation that we have considered are incompatible with this claim, for they all entail that experience of *all* colors has beaten science to every punch. However, this more general objection does not touch partial revelation.

A related objection is given in Byrne and Hilbert (2007). They observe that self-intimation entails a primitivist view of color (77). After all, colors do not seem on reflection to be essentially identical to anything specifiable in scientific terms (e.g., surface reflectance properties). They go on to suggest that primitivism about color is unsustainable. The partial revelationist avoids this objection, for they are not committed to self-intimation.

21. See Johnston (1992, 223).

22. See Allen (2011, 166).

23. See Campbell (2005, 111). Campbell in fact speaks of *categorical* properties, but it is intrinsic properties that he has in mind: see appendix 6A.

24. See Ivanov (2019).

25. Ivanov (2019, 158).

26. Perhaps *modulo* haecceitistic aspects (if haecceities are understood as intrinsic).

27. Ivanov borrows this account from Denby (2014).

28. There is a pressure point here: it is not immediately obvious why it will be revealed to the subject that purple is a *real* property. But perhaps the solution is to say that it is a priori that any intrinsic property is a real property.

29. Ivanov makes this explicit: "If [Full] Revelation is the case, experiencing such properties places us in a position to know exactly and fully what they are like. For example, supposing Purple is essentially compound in hue, experiencing Purple should put one in a position to know this" (Ivanov 2019, 147).

30. This view is defended throughout Pautz (2007). See especially p. 508.

31. Pautz does not mention essences, but this is an obvious way for him to account for the truth of partial revelation.

Chapter 3

1. It may be objected that there is some evidence, first, that cortically blind subjects cannot visually imagine anything and, second, that ordinary human subjects cannot vividly imagine specific colors. My reply is that both limitations are at best nomological. It is surely *metaphysically* possible for there to be cortically blind subjects who can vividly imagine specific colors, and that is all my arguments require.

2. See Mackie (2006).

3. See Kripke (1972, 110–115).

4. I thank Boyd Millar for urging me to consider putative essential truths like these.

5. This view is defended in Siegel (2006), Bayne (2009), Masrour (2011), Fish (2013), and McClelland (2016).

6. Although perceptions and episodic memories will turn out to involve very different kinds of successful sensory representations.

7. This portion of rich pluralism is especially similar to the representationalist theories developed in Tye (2009) and Schellenberg (2018).

8. I thank Ori Beck, Heather Logue, and Alex Moran for helping me appreciate the explanatory power of impure revelatory monism. A nearby view is held by Mark Johnston, who proposes that perception involves awareness of universals and particulars (2004, 135) and adds that experiential awareness of colors, and presumably other sensory qualities, is fully essence-revealing (1992, 223–224). The key difference is that Johnston accepts full revelation, while the impure revelatory monist accepts only partial revelation. Conduct (2012) develops a theory very similar to Johnston's, except that he does not commit to any version of revelation.

9. See Schellenberg (2018, 146–148) and Sethi (2020, 589).

10. This cluster of objections appears in Schellenberg (2018, 146). The first objection, pertaining to spatiotemporal extension, also appears in Pautz (2007, 517).

11. This is a variant on the well-known dualist view that it is just a brute law of nature, or metaphysics, that when certain neural states occur, there occurs the kind of representation that constitutes hard consciousness. See Chalmers (2004), Pautz (2010), and Mendelovici (2018).

12. This idea is closely related to the following proposal from Michael Tye:

> A color is visually represented in virtue of the fact that the visual system has been so designed (by Mother Nature) that when it is operating in the conditions it is supposed to operate in, a certain brain state in the visual cortex tracks or causally co-varies with a certain surface color (at least in first approximation). Under abnormal conditions, there need be no surface and the color may be uninstantiated. So, uninstantiated colors can be visually represented. (2019, 274)

But where Tye wants to explain the possibility of *representing* uninstantiated universals, I want to explain the possibility of being *aware* of them.

Similarly, Bengson, Grube, and Korman (2011) anticipate, at least in a schematic way, the possibility of an account like mine:

> We [do not] deny that there is *some* causal story to be told about why it is that a subject is aware of a given color when that color is not instantiated by any perceived worldly object.

> We deny only that the story will be so crude as to require that *the property itself* be causally responsible for the subject's state of awareness. (173)

13. It is less clear to me that a particular must be *spatiotemporally extended* to be a target of experiential awareness. Why couldn't a creature with the right perceptual apparatus experience a spacetime point?

14. See Pautz (2007, 517–518).

15. I agree with much of what is said in Tye (2019), but on this point, we part ways. Tye agrees that a visually hallucinating subject is aware of certain uninstantiated properties, but he thinks that the subject *sees* these properties (2019, 272).

16. I take this response directly from Bengson, Grube, and Korman (2011, fn. 39).

Chapter 4

1. In my (2019), I discuss hard problems concerning consciousness, normativity, and the self.

2. See Schaffer (2016) and A. Wilson (2018). Some will skeptical of the very notion of grounding—see, for example, J. Wilson (2014)—but I trust that my ideas could be expressed in other terms.

3. Byrne (2006) does a fine job of making this point vivid.

4. This is a central line of thought in Schaffer (2017).

5. The differences between CQ1 and CQ2, and the prima facie possibility that only one of them might constitute a hard problem, are brought out clearly in McClelland (2014).

6. This is the view of William Fish (2008). I have reproduced Fish's definition of acquaintance from p. 173. Fish adds: "the 'what' it's like for the subject is a matter of which mind-independent property the subject is acquainted with: the greenishness of a green experience just is the real-world instance of greenness that the subject is acquainted with . . . [In addition,] if the right kinds of functional processing are taking place, then the system will thereby be acquainted with aspects of its environment" (174). Fish claims on p. 174, and repeats on p. 175, that this provides the kind of "intelligible, transparent explanation" that dissolves the hard problems of consciousness.

7. On this last point, I am in perfect agreement with Pautz (2013, 28).

8. This is the view of Alex Byrne (2006): "This paper has two conclusions. First, Shoemaker is right: there is a 'hard problem' of color. Second, once we recognize the source of the puzzlement, *the mind-body problem disappears*. There may an unbridgeable gulf between color and ways of reflecting light; there is none between consciousness and brain process" (243, italics added). I interpret the italicized phrase

as saying that the mind–body problem *reduces* to the hard problem of color (or of sensory qualities more generally).

9. I thank Brian Cutter for an enlightening discussion of this point.

10. This is the view of Tye (2009). He begins by saying, "I am acquainted with the color scarlet, the city of Athens, the Apple computer at which I am now typing, the feeling of pain, the urge to gamble a large sum of money, and the feeling of jealousy. I have encountered (or am now encountering) all these things in experience" (2009, 101). Thus, he accepts C1. In his section on the explanatory gap problem, Tye adds, "Our sense that something significant is left out from our account of the underlying physical facts can be explained at least in part by means of the distinction between knowledge by description and knowledge by acquaintance" (Tye 2009, 139). As far as I can see, this idea is best expressed in terms of C2 and C3.

11. For further discussion of how cases of this sort bear on the idea that we experience high-level properties, see Byrne (2009, 449–450), Pautz (2009, 506), and Price (2009, section 5).

12. Questions such as this are of focal interest for Campbell (2002, chaps. 6–7).

13. William Fish argues that there are hard problems of consciousness constituted by questions like CQ10 (2008, 176–177). Umrao Sethi and Brian Cutter have suggested to me that the same might be true of questions like CQ11.

14. I was once confident in the truth of physicalism—see my (2019)—but I have become more impressed by the depth of the hard problems. I see compelling arguments both for and against physicalism, and I cannot find any good resolution.

15. On the explanatory gap problem, see Levine (1983). On the knowledge problem, see Nagel (1974) and Jackson (1982). On the conceivable absence problem, see Block (1980) and Chalmers (1996). On the conceivable inversion problem, see Shoemaker (1982). There may be other problems worth adding to this list: see Chalmers (2018).

16. On the notion of a natural kind, see Lewis (1983). In defining *hard character* as a natural kind term, I am taking inspiration from Lewis (1970).

17. For example, some properties might generate the knowledge problem but not the conceivable absence problem. See Carruthers and Veillet (2011), McClelland (2016), and Bayne (2022).

18. I am indebted to Keith Allen, Brian Cutter, Heather Logue, and Umrao Sethi for a fruitful discussion of these concerns.

19. This worry is developed in Campbell and Cassam (2014, 41–42).

20. This worry is expressed in Kennedy (2009, 577–578), Campbell and Cassam (2014, 41–42), and Genone (2016, 13).

21. This argument is due to Matthew Kennedy:

> The concept of phenomenal character has descriptive weaknesses. . . . When I see the green vase in my hallway, its manifest presence to me is not well captured by construing this presence as a property of my experience. This construal moves things in the wrong direction. By shifting subjective weight from the vase over to my experience of it, the concept of phenomenal character distorts the objectual presence that the notion of manifest presence tries to capture. This descriptive problem arises for our awareness of objects, and of properties. Here I assume that redness and other color properties are not properties of our experiences. When I see a 'Victory Red' Hummer vehicle, describing the ingredient manifest presence of redness to me in terms of the phenomenal character of my experience gives credit to the wrong actor. The concept of phenomenal character distorts the presence of objects and properties by converting this presence into a property of something else. (2009, 577–578)

22. This argument is due to James Genone:

> It is necessary to adopt a neutral formulation of what is meant by "perceptual phenomenology," namely the phenomena consciously encountered in perception. This characterization is distinct from the gloss usually provided when discussing *phenomenal character*, in terms of "what experience is like." What experience is like is a property of experience. . . . According to naïve realism, however, given that perceptual experience really does provide unmediated awareness of mind-independent objects and properties, the fact that it strikes us as so doing is best explained by the presence of the objects and their properties, not by a property of experience. In other words, claiming that the phenomenology associated with transparency is a property of experience is at odds with the phenomenon itself, which purports to involve the immediate presence of mind-independent objects. What this line of thinking suggests is that the notion of phenomenal character, construed as the property of what experience is like, is descriptively flawed when it comes to characterizing the phenomenology of perception. According to naïve realism, the fact that mind-independent objects and properties are experienced as immediately present is explained by the fact that they are constitutive of perceptual experience. This implies that at least some aspects of perceptual phenomenology are not adequately described in terms of the traditional notion of phenomenal character. (2016, 13)

Chapter 5

1. This definition is supplied by Pautz (2009, 492).

2. This definition appears in Siegel (2010, esp. 28).

3. Siegel is explicit on this point (Siegel 2010, 68–69).

4. Siegel welcomes this result: she wants naïve realists to be able to accept that experiences have contents (Siegel 2010, 73–76).

5. As Siegel (2010, 68) observes.

6. See, for instance, Tye (1995, 137–144), Burge (2010, chap. 9), and Schellenberg (2018, chap. 4, section 3).

7. Such cases have their origin in Grice (1961).

8. Some will object that we cannot imagine this kind of alien mental event from the inside in any detail. I agree, but I deny that this is a reason to think that such a mental event is metaphysically impossible: neither can we imagine what it is like to experience magnetoreception.

9. The distinction was first drawn in Block (1995), although he refers to hard consciousness as "phenomenal consciousness." I avoid the latter term because it is reminiscent of the expression "phenomenal character," which I find problematic: see chapter 10.

10. See Milner and Goodale (2008).

11. It has been argued that blindsight involves degraded but hard-conscious perception—see Phillips (2020). In that case, super-blindsight might be metaphysically impossible. Still, the example of simple creatures suffices to make the point.

12. It is usually thought that hard consciousness requires access consciousness but not vice versa. That is why I do not mention the possibility of perceptions that are hard-conscious but access-unconscious.

13. See Martin (2004, 2006).

14. See Siegel (2004, section 2).

15. See Tye (2009, chap. 4) and Schellenberg (2018, chap. 4).

16. It is worth comparing my arguments to those of Berger and Nanay (2016). They argue that naïve realists cannot account for hard-unconscious perceptions at all; I argue that impure revelatory monists (including impure revelatory naïve realists) *can* account for hard-unconscious perceptions, but only by adding a particular substantial posit to their theory.

17. For the first proposal, see Tye (1995, 138). For the second proposal, see Lycan (1996, chaps. 2–3) and Rosenthal (2005).

18. For what it is worth, I favor this last option.

19. I owe this objection to an anonymous referee.

Chapter 6

1. This restriction is also explicitly noted in Logue (2012a, 229).

2. Again, I am thinking of *grounding* as a metaphysical relation that backs metaphysical explanations in much the way that causation backs physical explanations—see Schaffer (2016) and A. Wilson (2018). Its relata are facts, and for p to (partly/fully) ground q is for q to obtain (partly/wholly) in virtue of p.

3. Campbell (2002) consistently elides the distinction. For example, he holds that "experience is experience of the categorical," but he calls this the *Intrinsicness Condition* (137). Similarly, he says, "We would ordinarily regard shape properties as the paradigmatic *categorical* properties of objects. Roundness is the reason why the thing tends to roll if suitably propelled and so on; it is the *intrinsic* ground of the complex functional state described" (139, emphasis added). For a discussion of why Campbell

might make this mistake, as well as an attempt to interpret Campbell's ideas in a charitable way, see appendix 6A.

4. Campbell identifies the categoricity datum, though he runs it together with the intrinsicality datum: see fn. 2.

5. This is a central posit of Campbell (2002, especially chaps. 6–7). See also Johnston (2006, 284–285), Logue (2012a, 229), and Ivanov (2019, section 4).

6. So says Campbell: "Suppose, however, that we drop the notion that the intrinsic characteristics of experience are *caused* by the categorical objects and properties being seen. Suppose instead we think of the intrinsic characteristics of experience as *constituted* by the categorical objects and properties being seen. . . . Then we are indeed in a position to understand how the phenomenal content of experience can explain our grasp of what the world is intrinsically like" (2002, 156).

7. Martin adopts a version of this view that is restricted to hallucinations that causally match veridical perceptions: "the disjunctivist insists that there is only a negative characterization of causally matching hallucinatory experience: it is nothing but a situation which could not be told apart from veridical perception . . . this is then the fundamental mental character of the event" (2004, 72). Martin's restricted view is equally subject to the criticism that I will press.

8. In a similar spirit, it has been argued that the indiscriminability view makes it impossible to explain other data—for example, that hallucinating red positions you to acquire knowledge about redness (Johnston 2004) or at least to form beliefs about redness (Pautz 2007, 525) and that hallucinating a butterfly positions you to know that you are not veridically perceiving a sausage (Siegel 2008, 211–214). I am tentatively sympathetic to these arguments.

9. This style of response is recommended in Soteriou (2016, 180).

10. Though she could learn this through other modalities like touch, we may suppose that, in fact, she has not done so.

11. I put this term in scare quotes for reasons explained in chapter 10. I imagine that this is intended to refer to either *hard character* or what I will later call *strong accessible character*, although I am not sure.

12. See Fish (2009, 94–95).

13. This view is defended in Alston (1999, 191–192). Naïve realist Alex Moran tells me that he accepts it, as well.

14. Flexible naïve realism is developed in Sethi (2020, 2021a, 2021b).

15. The imaginative view of hallucination is defended carefully in Allen (2015). On pp. 295–296, Allen mentions some examples of involuntary sensory imaginings that

are not hallucinations. He makes no attempt to complete the account by offering a full characterization of the nature of hallucination.

16. This view of sensory imagining is defended in Martin (2001, 270–273) and (2002, 404, 407). Allen entertains this view without committing to it (2015, 289).

17. See Johnston (2004) and Conduct (2012).

18. The argument is from Campbell (2002, 145–153). The core of his reasoning is this (with labels added for some key claims):

> All we have, to differentiate between [dispositional and categorical representations], is what the subject makes of the representation. Does the subject interpret the representation as relating to the dispositional or as relating to the categorical? (A2) All that we have to appeal to, as constituting the subject's interpretation of the representation in one way or another, is the pattern of functional relations in which the representation stands to other representations and ultimately the actions of the subject. But it is very hard to see how the existence of such a pattern of functional connections could constitute the subject's interpreting the representation as relating to the categorical, rather than the dispositional, characteristics of the external stimuli. For the pattern of functional connections is determined only by the dispositional characteristics of the external stimuli. (A4) If you varied the intrinsic nature of the objects around the subject, but kept their dispositional characteristics constant, the very same pattern of functional connections would still be appropriate for the use of the subject's perceptual representations. And the actions of the subject would also be of the same type, in responding to and manipulating the affordances of the external stimuli. So (A5 repeated) on a Representationalist view, we seem to have no motive for going beyond [dispositional representations], to suppose that the subject is representing the categorical. (Campbell 2002, 150)

Campbell adds:

> (A3) It is hard to see how brain states could be causally sensitive to anything more than the dispositional. For if we held the dispositional constant, but varied the categorical, the brain would still respond in the very same way. (Campbell 2002, 151)

19. Here I am mildly elaborating on the case from Campbell (2002, 149).

20. Campbell concedes this point:

> Of course the argument is not conclusive, since the categorical objects and properties are playing a role in the causation of our states, just because they underlie the dispositional properties; *but the causal theory of content itself is unlikely to provide the rationale for resistance.* (2002, 151)

My interpretation of the italicized phrase comes in the next paragraph.

21. For the idea that content is partly determined by naturalness, see Lewis (1983) and Sider (2011, esp. chap. 3).

22. Primitivist representationalism is developed in Chalmers (2004, section 6) and Pautz (2020).

Chapter 7

1. It is Bertrand Russell (1911) who seminally argues that experience positions us to know what we are referring to. A similar point appears in Johnston (2006, 264–265), but the idea has been given its fullest expression by John Campbell (2002, chaps. 6–7). My example is a mild variant of one that Campbell provides on p. 115.

2. See Campbell (2002) and Johnston (2006).

3. See Campbell (2002, 134).

4. Here I take inspiration from Mark Johnston, who observes that hallucinations position the subject to make *de re* reference, for the first time, to sensory qualities (2004, 130–131). That said, it is not entirely clear to me whether Johnston is discussing the special kind of *knowledge* of reference that is required for strong referential character.

5. Schellenberg (2018, esp. 126–128) offers this account as an explicit response to Campbell. I suspect that the account will be attractive to other singular representationalists, but it is hard to say, since many of them do not consider the matter of strong referential character at all. A prime example of this is Tye, whose 2009 book does not even include a citation of Campbell.

6. I press this complaint against Schellenberg. When discussing Campbell, she thinks that his concern is that "representationalist views cannot adequately account for the fact that perception grounds demonstrative thoughts and singular thoughts about particulars in the environment" (2018, 111). However, the deepest version of the concern is not about reference, but *knowledge* of reference. Later, Schellenberg does briefly mention knowledge of reference (124), but she continues to overlook the special kind of knowledge that is at issue.

7. This is a mildly refined version of Campbell's account. Campbell says,

> I have said that experience of the object is what provides knowledge of the reference of a demonstrative. . . . But we can further ask: what is it about experience of the object that allows it to play this causing and justifying role? (2002, 137)

By way of answer, Campbell appeals to "*The Intrinsicness Condition*," which says that "experience is experience of the categorical" (Campbell 2002). Notice that Campbell confuses intrinsicality with categoricity, as I noted in appendix 6A in chapter 6. In addition, Campbell's emphasis throughout his 2002 book is squarely on hard-conscious *perceptions*, as opposed to other kinds of experiences.

8. As we saw in the last chapter, this is the view of Campbell (2002).

9. As Campbell puts the point, "On the Representational View, . . . given a single background environment, you could have representational contents relating to daggers and their characteristics whether or not there actually was a dagger in front of you. So the phenomenal character of your experience could be exactly the same in

both cases" (Campbell 2002). I will argue in chapter 10 that Campbell is using the term *phenomenal character* to refer to strong referential character.

10. In Campbell's words:

> Experience is what explains our grasp of the concepts of objects. But if you think of experience as intentional, as merely one among many ways of grasping thoughts, you cannot allow it this explanatory role. . . . If all there is to experience of objects is the grasping of demonstrative thoughts about them, then experience of objects is just one among many ways in which you can exercise your conceptual skills. At this point we do not have any way of explaining why there should be anything fundamental to our grasp of concepts about experience of objects. (2002, 122)

11. For perspicuous descriptions of this phenomenon, though often under other names, see Alston (1999, 182), Sturgeon (2000, 9), and Millar (2014a, 240).

12. Here I am indebted to Millar (2014a, section 2), who offers an especially perspicuous description of the distinctive phenomenologies of perceptual experience and sensation. In addition, as Mackie (2020) emphasizes, a perceptually experienced maple tree will introspectively seem to exist independently of the experience *of the subject*. That is compatible with its existence depending on *other* minds, such as the mind of God.

13. The classical expression of the idea is Moore (1903, 450), although in the end he pulls back from endorsing a transparency thesis. Recent discussions of transparency trace to Harman (1990). The transparency thesis that I have stated is very close to the one stated in Tye (2002, 137–138).

14. On this point I disagree with Kennedy (2009). Kennedy is clearly discussing apparent presentational character (which he calls "manifest presence"), for on pp. 578–579, he approvingly cites Alston's description of that phenomenon. But Kennedy holds that transparency entails apparent presentational character (576–577, 579).

15. Some theorists—including some of those quoted above—use the term *awareness* to refer to the relation of presentation. I do not: hard-unconscious perceptions involve awareness, in my sense, without presenting anything. Even cases of abstract knowledge are cases of awareness, in my sense: see chapter 1. This issue is purely terminological.

16. For instance, Martin says that "[the naïve realist's] theory of appearance is put forward on the grounds that it gives an accurate description of how the subject's situation strikes her when consciously perceiving . . . in its very conscious and so subjective character, the experience [of seeing a lighthouse] seems literally to include the world" (1997, 4). Alston says that "the most intuitively attractive way of characterizing my state of consciousness as I [see a scene] is to say that it consists of the *presentation* of physical objects to consciousness. . . . The theory of appearing (hereinafter 'TA') is distinguished from rival theories by sticking close to this natural construal" (1999, 182). Fish adds that "naïve realism can be taken to have a certain

cachet as the theory that legitimates a pretheoretical, 'naïve' view of veridical visual experience" (2009, 18). In his subsequent defense of this idea against various objections, Fish clarifies that he has in mind a "kind of direct presence to the mind that the relation of acquaintance is intended to capture" (Fish 2009, 20). Insofar as Fish's discussion draws heavily on Hellie (2007a), the latter is a fourth example.

17. Fish (2009), although the account is not usually developed very explicitly. Logue (2012a) is a welcome exception. It is clear that Logue accepts IP1**. In addition, she suggests that the distinctive epistemic value of hard-conscious perception is that it "gives its subject *insight into what things in one's environment are like independently of one's experiences of them*" (227). She clarifies that "what I am suggesting is that the phenomenal character of veridical experiences of any Berkeleian property reveals *quite a bit* about its intrinsic nature (although not everything)" (229). Thus, she accepts IP2**.

18. Again, see Campbell (2002, especially chaps. 6–7), Johnston (2006, 284–285), Logue (2012a, 229), and Ivanov (2019, section 4).

19. It is plausible that the subject's perception must be veridical in some respect, as well: an experience that is illusory about *every* property of a perceived particular plausibly does not present that particular. I will suppress this minor qualification below.

20. For the first proposal, see Tye (1995, 138). For the second proposal, see Lycan (1996, chaps. 2–3) and Rosenthal (2005). For the third proposal, see Chalmers (2004) and Pautz (2020).

21. As Campbell puts it:

> On the Representational View, . . . all that is different between the two cases [of perception and hallucination] is the way in which the representational contents are caused on these occasions. In the veridical case, the representations are caused by a dagger. In the hallucinatory case, the representational contents are but creatures of the heat-oppressed brain.
>
> On the Relational View, in contrast, . . . [in] the case in which there is a dagger, the object itself is a constituent of your experience. The experience is quite different in the case of the hallucination, since there is no object to be a constituent of your experience. (2002, 117)

A similar concern appears in Logue (2012a, 232–235).

22. The broad possibility of such a view has been anticipated by Harold Langsam. Langsam is a naïve realist, but he concedes that "if there is a kind of representation that is a genuine relation, then for all that has been said, the relation that is of concern to the naïve realist can be a kind of representation" (2017, 111).

23. Campbell observes, "On the Relational View, in contrast [to the Representational View], there is nothing intrinsic in common between the cases in which there is a dagger to which you are consciously attending, and the case in which you are just having a hallucination" (Campbell 2002, 117). Similarly, Fish says, "the hillsides shape the contours of the landscape by actually *being* the contours of the landscape. This, I suggest, is how we should understand the naïve realist's claim that external objects and their properties shape the contours of the subject's conscious

experience: they shape the contours of the subject's conscious experience by actually *being* the contours of the subject's conscious experience."

24. See Campbell (2002, 116–117) and Logue (2012a, 232–235).

Chapter 8

1. It is hard to find explicit avowals of the easy account in the literature. However, I report that a number of philosophers of perception, when they have read this portion of the book, have expressed sympathy for the easy account.

2. See Perky (1910). For an alternative interpretation of these results, see Hopkins (2012, 2013); for a response, see Nanay (2012b).

3. This is the view of Adam Pautz:

> Why then do experiences, including hallucinatory experiences, have a presentational phenomenology while thoughts do not? . . . My answer to this question is that there is no answer. Some intentional states (believing contents, desiring contents) do not have a presentational phenomenology. Other intentional states do have a presentational phenomenology: states in which one stands in the special "sensory" relation to contents (propositions or complex properties) which I have dubbed 'sensorily entertaining'. (Or at least some episodes of standing in this sensory relation to contents have a presentational phenomenology; perhaps in imaginings and dreams we stand in the same relation to contents but *because the contents are quite indeterminate* the experiences lack a presentational phenomenology.) The state of having an experience as of the vivid presence of a red thing at a certain place just is the state of standing in this special relation to a *detailed* content according to which there is a red thing at a certain place. (2007, 519, emphases added)

4. A naïve realist cousin of this account appears in Dokic and Martin (2012).

5. Austin Andrews defends the related suggestion that as we mature, we form a theoretically laden concept of *perception* (see his 2020, esp. 236–239). I am sympathetic to that idea as well.

6. Though there are possible creatures that form and/or apply the same concept in very different ways.

7. A helpful philosophical overview of this phenomenon can be found in Dokic and Martin (2017, 301–302). Miyazono (2021) uses findings about derealization disorder to argue that not all perceptual experiences have apparent presentational character.

8. See, for example, Bayne (2009), Fish (2013), and McClelland (2016).

9. See, for example, Byrne (2009), Pautz (2010), and Carruthers and Veillet (2011).

10. Evans (1982) introduces the notion of nonconceptual content. See Heck (2000) for a discussion of how precisely this notion might be understood.

11. Since I analyze apparent presentational character in terms of the application of a certain concept, this claim would entail that experiences only ever have apparent

presentational character accidentally. Is this consequence problematic? No, it is independently plausible. Take a typical conscious perception with apparent presentational character. If that experience were had by an animal that did not possess any concepts, then it would not seem to present anything to that animal. Still, it *would* be essential to this experience to have the gestalt that it does, so the *basis* for apparent presentational character would remain.

12. This evidence has been aptly summarized by Mandelbaum (2018). I will follow his presentation closely.

13. See Potter et al. (2014).

14. See Mandelbaum (2018, 268).

15. Mandelbaum (2018, 275).

16. Mandelbaum (2018).

17. This claim might be contested by, for instance, higher-order representationalists, and some of them will say that it is prima facie plausible that any experience involves awareness of that very experience. But this suggestion has received a very mixed reception. Thus, it is reasonable to deny it in the absence of further arguments.

18. As Campbell puts it, "On the [representationalist theory], all that experience of the object provides you with is a conscious image of the object—the image which bears the representational content. The existence of that conscious image is in principle independent of the existence of the external object. The existence of the image, though, is dependent on the existence of the subject who has the conscious image. So if your conception of the object was provided by your experience of the object, you would presumably end by concluding that the object would not have existed had you not existed, and that the object exists only when you are experiencing it. We cannot extract the conception of a mind-independent world from a mind-dependent image" (2002, 121).

19. Although the *higher-order representationalist theory* holds that the subject perceptually represents *both* mind-independent entities *and* their own perceptual representations. See, for example, Lycan (1996, chap. 2) and Rosenthal (2005).

20. This is an alternative way of interpreting Campbell's remarks cited in the footnotes of section 7.4 in chapter 7. This interpretation is driven by the fact that, at times, Campbell seems to emphasize not just our grasp of objects themselves but of their mind-independence. For example, at one point, Campbell says, "The objection to the common factor view is that on it, experience of objects could not be what explains our having the conception of objects as mind-independent" (2002, 121). Here Campbell is stating an objection taken from McDowell, but he goes on to develop McDowell's objection himself.

21. It is important to include the restriction to beliefs that are contingent and a posteriori. Necessarily true beliefs might trivially turn out to be safe and sensitive. Likewise for some contingent but a priori true beliefs, such as the belief that I exist.

In addition, if the case is unusual enough, my belief that I am presented with a mind-independent ball might fail to be safe or sensitive, and it might thus fail to constitute knowledge. For instance, perhaps I have taken a pill that causes me to hallucinate half the time, but I happen not to be hallucinating now. Perhaps I even know this.

22. Perhaps the content is singular with respect to subject and time; perhaps not. It will not matter.

23. The seminal expression of this account comes from John Searle. Searle begins by identifying apparent presentational character as one of his explanatory targets:

> If, for example, I see a yellow station wagon in front of me, the experience I have is directly of the object. It doesn't just "represent" the object, it provides direct access to it. The experience has a kind of directness, immediacy and involuntariness which is not shared by a belief I might have about the object in its absence. (1983, 45–46)

Searle then lays out his explanatory apparatus:

> For example, if I see the yellow station wagon, I have a certain visual experience. But the Intentional content of the visual experience, which requires that there be a yellow station wagon in front of me in order that it be satisfied, also requires that the fact that there is a yellow station wagon in front of me must be the cause of that very visual experience. (1983, 47–48)

The causal content account receives further development in Millar (2014a, 2014b) and Schroer (2014).

24. See, for example, Millar (2014b, 649–650). Some points from Siegel (2006, section 5) would also apply, though she is not defending a causal content theory there.

25. In principle, it is possible to couple the causal content account with *singular* rather than existential representationalism. Making this change to the account would require making many other changes, as well. I invite the causal content theorist to try this out. For what it is worth, I suspect that many complications would need to be introduced to make the theory adequate, leaving the categorization account as the simpler option.

26. This view is defended at length in Siegel (2006).

27. Siegel (2006, 355).

28. Siegel (2006, 369–370).

29. Siegel makes it explicit that the phenomenological difference between typical visual perceptions and typical sensations is her explanatory target. As she puts things in the opening lines of her (2006) paper,

> What is the difference between perception and mere sensation? Take a typical perceptual experience, such as an experience of seeing a fish or a table, and a merely sensory experience, such

> as the experience of "seeing stars" or of enjoying a red phosphene. One difference between these experiences is that in the first case there is an external object that one sees. But this difference is not the only difference. On the face of it, typical perceptual experiences and mere sensations also differ in their phenomenal character. How can this difference be understood?
>
> In this essay, I will argue that there is a *representational* difference between perceptual experiences and mere sensations.

Note especially Siegel's use of the word *is* in her very first sentence. This makes it clear that her explanatory target is not what *causes* the difference between perception and sensation but what *constitutes* that difference.

30. Siegel is explicit on this point (2009, 360).

31. Often, but not always. Remember the example of sky-perceptions and sky-afterimages.

32. *Contra* Fish (2008).

Chapter 9

1. Flexible naïve realism is carefully developed in a series of papers by Umrao Sethi: see her (2020, 2021a, 2021b).

2. Sethi (2020, 606–609).

3. This is the view defended in Allen (2015).

4. This view is defended in Martin (2001, 270–273) and (2002, 404, 407). Allen (2015, 289) entertains the view without committing to it.

5. This is Allen's preferred move. See his (2015, 299–301).

6. See Johnston (2004) and Conduct (2012).

7. As Johnston puts it:

> In the case [of seeing supersaturated red], the complex will be the property of being a supersaturated red round thing at a certain changing distance and direction from the present position, at the present time. The relational element—the spatio-temporal layout as I have called it—mimics spatial and temporal extent and thereby mimics particularity, so that although it seems as if an after-image is a moving particular, in fact there is just a complex of sensible qualities and relations. . . . Thanks to containing certain properties in certain relations to continuous places and times, a primary object can immediately strike the subject as a moving particular. (2004, 142)

8. For a nonpluralistic version of this naïve realist approach, see Martin (2002).

Chapter 10

1. Here I refine my presentation of the problem from my (2014, 311–312).

2. I include the parenthetical qualification to cover the view of, for example, Tye (2009). Tye thinks that the content of perception concerns particulars and qualities,

but he adds that only the qualities, not the particulars, are constituents of phenomenal character (119).

3. William Fish outlines this dilemma by considering a case in which Michael Tye sees the blueness of the Pacific Ocean:

> [A] representationalist might claim that veridical experiences possess a further content that is singular. While such a position would be possible, and may indeed allow for the claim that the very blueness of the ocean is *seen* by Tye (where seeing is a matter of veridical representation), as long as it is an abstract or general content that is identified with/determines phenomenal and hence presentational character, then the representationalist will be unable to hold that it is the very blueness of the ocean that Tye is *presented with*. And if a representationalist were to hold that this singular content determined phenomenal/presentational character, then such a phenomenal character would be unavailable in the nonveridical cases. (2009, 23, fn. 27)

4. One tricky example is the expression *I*. There is *some* specificity to what the expression *I* can refer to. But a lot is left open, as well: *I* can refer to something that is not a person (imagine a sign that says "I am written in English"). So is the context sensitivity of this expression deep or shallow? I do not know.

5. See Nagel (1974). His term is actually *subjective character*, and Nagel focuses in the first instance on what it's like to be a subject. None of this affects my point.

6. This definition is now so dominant that it would be pedestrian to give further citations.

7. See, for example, Snowdon (2010, 22), Stoljar (2016, 1174–1175), and Gaskin (2019, 679, 682).

8. I take this example from Gaskin (2019, 675).

9. The second and third of these answers are drawn from Wikipedia.

10. See Snowdon (2010) and Hill (2014, 230–231).

11. See, for example, Stoljar (2016, 1162).

12. For instance, Lormand (2004) and Stoljar (2016, esp. section 7).

13. See Lewis (1995, 140) and Byrne (2004, 215).

14. On the point that philosophical "what it's like" talk is not technical, I side with Farrell (2016).

15. These are the paradigmatic theories of phenomenal character offered by representationalists, naïve realists, and type-physicalists, respectively.

16. On the basis of claims like VPC1, and contrary to philosophical orthodoxy, Hellie argues that emotional states or properties *are* phenomenal characters (2007a, 261–262). But if the expression "what it's like" is deeply context sensitive, then

Hellie's reasoning is too swift: he has not ruled out the possibility of a shift in context.

17. See Gaskin (2019).

18. Gaskin (2019, 21).

19. See Hellie (2007b, 452).

20. See Gaskin (2019, 693).

21. See Snowdon (2010, 21–22).

22. See Snowdon (2010, 22) and Stoljar (2016, 1169–1172, including fn. 14).

23. See Stoljar (2016, fn. 14).

24. See Stoljar (2016, 1173). Stoljar's actual proposal is about the expression "there is something it's like . . ." rather than the expression "what it is like," but presumably this is how Stoljar's proposal would be extended.

25. See Lormand (2004, 315).

26. Lormand (2004, 316–317).

27. See Hellie (2007b, 443–444, 449–459), Stoljar (2016, 1186–1188), and Gaskin (2019, 676–680).

28. This practice is so ubiquitous that I trust that citations are unnecessary.

29. For example, Chalmers uses the "what it's like" definition of consciousness and says that "the subject matter [of a discussion of consciousness] is 'the *subjective* quality of experience'" (1996, 4, emphasis added); Tye equates phenomenal aspects with what-it-is-like aspects and adds that "a mental state is phenomenally conscious just in case there is some immediate *subjective* 'feel' to the state" (1995, 4, emphasis added); Sturgeon (2008) explains that "the conscious character of Good cases" is "what it is *consciously* like to enjoy Good visual experience" (p. 116, emphasis added).

30. Lycan (1996, chap. 1) identifies eight.

31. There may be further roles that also relate to the hard problems of consciousness. See Chalmers (2018).

32. See pp. 16–17, 12–13, 21–25, and 26–29, respectively. One minor complication is that Tye thinks of phenomenal character not as a *property* of an experience but as the *content* of the experience. If anything, this is grist for my mill: here is even more fragmentation in uses of "phenomenal character."

33. Dretske (1995), 81.

34. Dretske (1995, 81–95).

35. Campbell does mention inverted spectra in his (1993, 268). But this is brief, and more importantly, Campbell is not discussing phenomenal character here.

He has a more extended discussion of inverted spectra in chapter 2 of *Berkeley's Puzzle*. However, this book was published in 2014, a full twelve years after *Reference and Consciousness*. In light of this, and the evidence cited below, I conclude that the relevant section of *Berkeley's Puzzle* is discussing claims about phenomenal character that Campbell regards as synthetic rather than analytic.

36. The remarks from the previous footnote apply again.

37. A perceptive reader might have noticed that the expression "what it's like" does not appear in any of the passages from Campbell that I quoted in section 10.4. Perhaps, then, Campbell is simply defining the term *phenomenal character* directly, without appeal to what it's like to have an experience. Would this affect any of my central conclusions?

It would not. To begin with, the quotations would still show that when Campbell uses the term *phenomenal character*, it is strong referential character that he has in mind. We should continue to conclude that Campbell is talking past philosophers like Dretske and Tye.

In addition, the present concern still allows the deep context-sensitivity hypothesis to play a crucial explanatory role. For it is surprising that Campbell talks past Dretske and Tye at all. Campbell is plainly familiar with their work. Why then would he fail to understand how they are using the term *phenomenal character*? The deep context-sensitivity hypothesis suggests an answer. Campbell was immersed in certain philosophical problems about perception. Thus, when Campbell encountered the standard "what-it's-like" definition of phenomenal character, he assumed, mistakenly but quite reasonably, that his interlocutors were speaking in a context in which such problems were salient.

38. Campbell (1993, 4).

39. Campbell (1993, 51–52).

40. Campbell (1993, 44).

41. Siegel (2010, 51). Siegel also speaks of entities other than contents that can be conveyed to the subject—properties, for example—so presumably her remarks above can somehow be generalized to explain what it is for these other entities to be conveyed to the subject.

42. See Block (1995).

43. I happen to find this claim plausible. Some spatiotemporally located particles might have mass 0, but the universal red does not have any mass at all, not even mass 0.

44. Why *red* rather than *scarlet*? Because the content of a categorization will typically concern a merely determinable color.

45. This is the thesis of Schwitzgebel (2016).

46. Incidentally, I would argue that this is not the correct interpretation of Tye. In *Consciousness Revisited* (2009), Tye says that what it is to perceive is to deploy a representation of a certain kind. He argues at great length in chapter 5 that perceptual representations have singular contents: when I see a particular mango, *that very mango* is part of the content of my perception. He even says that "in the case of perceptual experience, what experiences *fundamentally* aim to do is put us in contact with objects around us" (Tye 2009, 85). Thus, Tye clearly thinks that my (hard-conscious) perception makes the particular mango directly accessible to me.

At the same time, Tye insists that "the phenomenal character of a perceptual experience consists in, and is no more than, the complex of qualities the experience represents. Thus, the phenomenal character of the experience of red just is red" (Tye 2009, 119). Evidently, Tye is not using the expression *phenomenal character* to refer to strong accessible character. His remarks make perfect sense if hard character is what he has in mind.

Chapter 11

1. See Martin (2004, 43), Soteriou (2005, 178), Crane (2006, 139), Neta (2008, 311–312), Fish (2009, 94–95), Nudds (2009, 337), Brewer (2011, 94), Hellie (2013, 151), Logue (2013, 109), Genone (2016, 7), and Gomes (2017, 534).

2. Naïve realist Michael Martin anticipates this argument against his own view (2004, 61–62; 2006, 369–372). Martin's preferred response is to reject B3 by accepting a deflationary view of what the common kind is (2004, section 7; 2006, 369–370).

Incidentally, Martin (2004, 62, fn. 17) attributes the argument to Sturgeon (1998, 2000, chap. 1), but I believe that Martin is too modest; the argument is his own. What Sturgeon says is rather that if the naïve realist offers different explanations of apparent presentational character ("Scene-Immediacy") for perceptions and hallucinations, then they will "thereby [be offering a] disunifying theory" (1998, 198). This is quite different from concerns about screening-off.

3. Relatedly, Benj Hellie discusses what he calls the *multimoderate view*: the view that the tomato-hallucination belongs to the fundamental kind defined by some common property *C*, while the tomato-perception belongs to the fundamental kind defined by the conjunctive property *being a visual perception of a tomato and being C*. This is subtly different from rich pluralism. For the multimoderate theorist buys into a certain standard presupposition—that any mental item belongs to exactly one *fundamental kind*—and says that for perceptions, this kind is conjunctive. For reasons that I explain in my (2022b), I prefer to drop this presupposition. Thus, I

say that the tomato-perception belongs to several distinct fundamental kinds, the conjunction of which is not itself a fundamental kind.

In any case, Hellie thinks that the screening-off argument succeeds against the multimoderate view (2013, 167). However, the response I will give for rich pluralism should also save this view.

4. So says Michael Martin (2004, 64) and (2006, 371). He presses this concern against any naïve realist who rejects B3, but his concern would apply equally to any rich pluralist who denies C3.

5. Martin endorses this argument. See especially his (2004, 59–60).

6. My opponent might want to add that this mental kind must be *fundamental*. This would not affect my response.

7. See Martin (2006, 369).

8. See Martin (2004, 53–58) and (2006, 369).

9. Here is how Martin expresses D4–D7:

> One might claim that, when one perceives, two events occur: there is the genuinely perceptual experiential event, and this has no analogue in hallucination; then there is the purely experiential event, which also occurs when one is hallucinating. Given this model it is both true that something occurs only when one perceives, the genuinely perceptual experience, and something occurs which can also occur when one hallucinates, the purely experiential event. But is this a plausible model?
>
> . . . What . . . of the relation between the supposed genuinely perceptual experiential event and the purely experiential event? If we ask of what happened in the subject's life such that they could reflect on and report it as such, then there seems to be just one thing to talk about. . . . In the [case of a subject who sees a lavender bush], there are supposedly two events to introspect: the purely experiential event and the genuinely perceptual one. So, although there seems to be some questions to be settled anyway about the ways in which to count experiences, this proposal seems to double for us the number with which we answer that question.
>
> If we say that in fact one can introspect both events in the perceptual case, then we need to explain why it should seem as if one cannot selectively attend to the one experience as opposed to the other. (2004, 59–60)

Chapter 12

1. For the sake of completeness, part III also considered strange pluralism. However, this view turned out to be unmotivated and plainly inferior to all other pluralistic options, so I will not discuss it any further in this chapter.

2. The rich pluralist also posits taking-as, of which categorization is a sophisticated form. But these posits are not *distinctive*: all theorists will need to posit such things.

3. This conclusion is of course provisional: it remains possible to construct new theories. For instance, Lisa Miracchi has developed a *perception-first theory* (2017), and

Justin D'Ambrosio has developed a novel version of adverbialism (2019). I admire the ingenuity of these theories. As they stand, however, I do not see how either one would account for the data described in chapter 2. Still, it is possible to construct pluralistic versions of these theories, which I imagine would fare better. One intriguing option would be to replace the rich pluralist's theory of sensory representation with Miracchi's theory of capacities to perceive, while preserving the theory of deep awareness as is. I regard this as an interesting challenger, one that could draw on almost exactly the same explanatory structures as rich pluralism itself.

References

Allen, K. 2011. "Revelation and the nature of colour." *Dialectica* 65 (2): 153–176.

Allen, K. 2015. "Hallucination and imagination." *Australasian Journal of Philosophy* 93 (2): 287–302.

Allen, K. 2020. "The value of perception." *Philosophy and Phenomenological Research* 100 (3): 633–656.

Alston, W. 1999. "Back to the theory of appearing." *Philosophical Perspectives* 13: 181–203.

Andrews, A. 2020. "Acquiring a concept of visual experience." *Philosophical Quarterly* 70 (279): 223–245.

Bayne, T. 2009. "Perception and the reach of phenomenal content." *Philosophical Quarterly* 59 (236): 385–404.

Bayne, T., and M. Montague. 2011. *Cognitive phenomenology*. Oxford: Oxford University Press.

Bayne, T. 2022. "The puzzle of cognitive phenomenology." In *Oxford studies in philosophy of mind*, vol. 1, ed. Uriah Kriegel. Oxford, UK: Oxford University Press.

Beck, O. 2019. "Rethinking naïve realism." *Philosophical Studies* 176: 607–633.

Bengson, J., E. Grube, and D. Korman. 2011. "A new framework for conceptualism." *Noûs* 45 (1): 167–189.

Berger, J., and B. Nanay. 2016. "Relationalism and unconscious perception." *Analysis* 76 (4): 426–433.

Block, N. 1980. "Troubles with functionalism." In *Readings in the philosophy of psychology*, vol. 1, ed. N. Block, 268–305. Cambridge, MA: Harvard University Press.

Block, N. 1995. "On a confusion about a function of consciousness." *Behavioral and Brain Sciences* 18: 227–287.

Block, N. 1996. "Mental paint and mental latex." *Philosophical Issues* 7: 19–49.

Brewer, B. 2011. *Perception and its objects*. Oxford: Oxford University Press.

Broad, C. 1952. "Some elementary reflexions on sense-perception." *Philosophy* 27: 3–17.

Broi, A. 2020. "Revelation and phenomenal relations." *Philosophical Quarterly* 70 (278): 22–42.

Burge, T. 2010. *Origins of objectivity*. Oxford: Clarendon Press.

Byrne, A. 2004. "What phenomenal consciousness is like." In *Higher-order theories of consciousness: An anthology*, ed. R. Gennaro, 203–226. Amsterdam: John Benjamins.

Byrne, A. 2006. "Color and the mind-body problem." *Dialectica* 60 (3): 223–244.

Byrne, A. 2009. "Experience and content." *Philosophical Quarterly* 59 (236): 429–451.

Byrne, A., and D. Hilbert. 2007. "Color primitivism." *Erkenntnis* 66: 73–105.

Campbell, J. 1993. "A simple view of colour." In *Reality: Representation and projection*, ed. J. Haldane and C. Wright, 257–268. Oxford: Oxford University Press).

Campbell, J. 2002. *Reference and consciousness*. Oxford: Oxford University Press.

Campbell, J. 2005. "Transparency versus revelation in color perception." *Philosophical Topics* 33 (1): 105–115.

Campbell, J., and Q. Cassam. 2014. *Berkeley's puzzle*. Oxford: Oxford University Press.

Carruthers, P., and B. Veillet. 2011. "The case against cognitive phenomenology." In *Cognitive phenomenology*, ed. T. Bayne and M. Montague, 35–56. Oxford: Oxford University Press.

Chalmers, D. 1996. *The conscious mind*. New York: Oxford University Press.

Chalmers, D. 2004. "The representational character of experience." In *The future for philosophy*, ed. B. Leiter, 153–181. Oxford: Oxford University Press.

Chalmers, D. 2018. "The meta-problem of consciousness." *Journal of Consciousness Studies* 29 (9–10): 6–61.

Conduct, M. 2012. "Naïve realism without disjunctivism about experience." *Consciousness and Cognition* 21 (2): 727–736.

Correia, F., and A. Skiles. 2019. "Grounding, essence, and identity." *Philosophy and Phenomenological Research* 98 (3): 642–670.

Crane, T. 2006. "Is there a perceptual relation?" In *Perceptual experience*, ed. T. Gendler and J. Hawthorne, 126–146. Oxford: Oxford University Press.

D'Ambrosio, J. 2019. "A new perceptual adverbialism." *Journal of Philosophy* 116 (8): 413–446.

Denby, D. 2014. "Essence and intrinsicality." In *Companion to intrinsic properties*, ed. R. Francescotti. Berlin: De Gruyter.

Dokic, J., and J. Martin. 2012. "Disjunctivism, hallucinations, and metacognition." *WIREs Cognitive Science* 3: 533–543.

Dretske, F. 1995. *Naturalizing the mind*. Cambridge, MA: MIT Press.

Evans, G. 1982. *The varieties of reference*. New York: Oxford University Press.

Farrell, J. 2016. "'What it is like' talk is not technical talk." *Journal of Consciousness Studies* 23 (9–10): 50–65.

Fine, K. 1994. "Essence and modality." *Philosophical Perspectives* 8: 1–16.

Fish, W. 2008. "Relationalism and the problems of consciousness." *Teorema* 27 (3): 167–180.

Fish, W. 2009. *Perception, hallucination, and illusion*. Oxford: Oxford University Press.

Fish, W. 2013. "High-level properties and visual experience." *Philosophical Studies* 162: 43–55.

Gaskin, R. 2019. "A defence of the resemblance meaning of 'What it's like'." *Mind* 128 (511): 673–698.

Genone, J. 2016. "Recent work on naïve realism." *American Philosophical Quarterly* 53 (1): 1–24.

Goff, P. 2017. *Consciousness and fundamental reality*. Oxford: Oxford University Press.

Gomes, A. 2017. "Naïve realism in Kantian phrase." *Mind* 126: 529–578.

Grice, H. 1961. "The causal theory of perception." *Proceedings of the Aristotelian Society* 35: 121–153.

Harman, G. 1990. "The intrinsic quality of experience." *Philosophical Perspectives* 4: 31–52.

Heck, R. 2000. "Nonconceptual content and the space of reasons." *Philosophical Review* 109: 483–523.

Hellie, B. 2007a. "Factive phenomenal characters." *Philosophical Perspectives* 21: 259–306.

Hellie, B. 2007b. "'There's something it's like' and the structure of consciousness." *Philosophical Review* 116: 441–463.

Hellie, B. 2013. "The multidisjunctive conception of hallucination." In *Hallucination*, ed. F. Macpherson and D. Platchias, 149–173. Cambridge, MA: MIT Press.

Hill, C. 2014. *Meaning, mind, and knowledge*. Oxford: Oxford University Press.

Hopkins, R. 2012. "What Perky did not show." *Analysis* 72: 431–439.

Hopkins, R. 2013. "Perky, phenomenal similarity and photographs: Reply to Nanay." *Analysis* 73 (1): 77–80.

Ivanov, I. 2019. "How to tell essence (when you sense one)." *Canadian Journal of Philosophy* 49 (2): 147–168.

Jackson, F. 1982. "Epiphenomenal qualia." *Philosophical Quarterly* 32: 127–36.

Johnston, M. 1992. "How to speak of the colors." *Philosophical Studies* 68: 221–263.

Johnston, M. 2004. "The obscure object of hallucination." *Philosophical Studies* 120: 113–183.

Johnston, M. 2006. "Better than mere knowledge? The function of sensory awareness." In *Perceptual experience,* ed. T. Gendler and J. Hawthorne, 260–290. Oxford: Oxford University Press.

Kappes, Y. 2020. "Phenomenal relations and collective essence." *Philosophical Quarterly* 70 (281): 800–808.

Kennedy, M. 2009. "Heirs of nothing: The implications of transparency." *Philosophy and Phenomenological Research* 79 (3):574–604.

Kripke, S. 1972. *Naming and necessity*. Cambridge, MA: Harvard University Press.

Langsam, H. 2017. "The intuitive case for naïve realism." *Philosophical Explorations* 20 (1): 106–122.

Leslie, S. 2011. "Essence, plenitude, and paradox." *Philosophical Perspectives* 25: 277–296.

Levin, J. 2007. "What is a phenomenal concept?" In *Phenomenal concepts and phenomenal knowledge,* ed. T. Alter and S. Walter, 87–110. New York: Oxford University Press.

Levine, J. 1983. "Materialism and qualia: The explanatory gap." *Pacific Philosophical Quarterly* 64: 354–361.

Levine, J. 2001. *Purple haze: The puzzle of consciousness*. New York: Oxford University Press.

Levine, J. 2006. "Conscious awareness and self-representation." In *Self-representational approaches to consciousness*, ed. U. Kriegel and K. Williford. Cambridge, MA: MIT Press.

Lewis, D. 1970. "How to define theoretical terms." *Journal of Philosophy* 67 (13): 427–446.

Lewis, D. 1983. "New work for a theory of universals." *Australasian Journal of Philosophy* 61 (4): 343–377.

Lewis, D. 1995. "Should a materialist believe in qualia?" *Australasian Journal of Philosophy* 73 (1): 140–144.

Logue, H. 2012. "Why naïve realism?" *Proceedings of the Aristotelian Society* 112 (2): 211–237.

Logue, H. 2013. "Good news for the disjunctivist about the bad cases." *Philosophy and Phenomenological Research* 86 (1): 105–133.

Lormand, E. 2004. "The explanatory stopgap." *Philosophical Review* 113: 303–357.

Lycan, W. 1996. *Consciousness and experience.* Cambridge, MA: MIT Press.

Mackie, P. 2006. *How things might have been: Individuals, kinds, and essential properties.* Oxford: Oxford University Press.

Mackie, P. 2020. "Perception, mind-independence, and Berkeley." *Australasian Journal of Philosophy* 98 (3): 449–464.

Mandelbaum, E. 2018. "Seeing and conceptualizing: Modularity and the shallow contents of perception." *Philosophy and Phenomenological Research* 97 (2): 267–283.

Martin, M. 1997. "The reality of appearances." In *Thought and ontology*, ed. M. Sainsbury. Milan, Italy: Franco Angeli.

Martin, M. 1998. "Setting things before the mind." *Royal Institute of Philosophy Supplement* 43: 157–179.

Martin, M. 2001. "Out of the past: Episodic recall as retained acquaintance." In *Time and memory*, ed. C. Hoerl and T. McCormack, 257–284. Oxford: Oxford University Press.

Martin, M. 2002. "The transparency of experience." *Mind and Language* 17 (4): 376–425.

Martin, M. 2004. "The limits of self-awareness." *Philosophical Studies* 120: 37–89.

Martin, M. 2006. "On being alienated." In *Perceptual experience*, ed. T. Gendler and J. Hawthorne, 354–410. Oxford: Oxford University Press.

Masrour, F. 2011. "Is perceptual phenomenology thin?" *Philosophy and Phenomenological Research* 83 (2): 366–397.

McClelland, T. 2014. "The problem of consciousness: Easy, hard, or tricky?" *Topoi* 2017 (36): 17–30.

McClelland, T. 2016. "Gappiness and the case for liberalism about phenomenal properties." *Philosophical Quarterly* 66 (264): 536–558.

McLaughlin, B. 2003. "The place of color in nature." In *Colour perception: Mind and the physical world*, ed. R. Mausfeld and D. Heyer, 475–502. New York: Oxford University Press.

Mehta, N. 2014. "The limited role of particulars in phenomenal experience." *Journal of Philosophy* 111 (6): 311–331.

Mehta, N. 2019. "Phenomenal, normative, and other explanatory gaps: A general diagnosis." *Philosophy and Phenomenological Research* 98 (3): 567–591.

Mendelovici, A. 2018. *The phenomenal basis of intentionality*. New York: Oxford University Press.

Millar, B. 2014a. "The phenomenological directness of perceptual experience." *Philosophical Studies* 170: 235–253.

Millar, B. 2014b. "The phenomenological problem of perception." *Philosophy and Phenomenological Research* 88 (3): 625–654.

Millar, B. 2017. "Thinking with sensations." *Journal of Philosophy* 114 (3): 134–154.

Milner, A. D., and M. Goodale. 2008. "Two visual systems re-viewed." *Neuropsychologia* 46: 774–785.

Miracchi, L. 2017. "Perception first." *Journal of Philosophy* 114 (12): 629–677.

Miyazono, K. 2021. "Visual experiences without presentational phenomenology." *Ergo* 8 (19): 551–576.

Moore, G. E. 1903. "The refutation of idealism." *Mind* 12: 433–453.

Nagel, T. 1974. "What is it like to be a bat?" *Philosophical Review* 83 (4): 435–450.

Nanay, B. 2012a. "Perceiving tropes." *Erkenntnis* 77 (1): 1–14.

Nanay, B. 2012b. "The philosophical implications of the Perky experiments: Reply to Hopkins." *Analysis* 72 (3): 439–443.

Neta, R. 2008. "In defence of disjunctivism." In *Disjunctivism: Perception, action, knowledge*, ed. A. Haddock and F. Macpherson, 311–329. Oxford: Oxford University Press.

Nudds, M. 2009. "Recent work in perception: Naïve realism and its opponents." *Analysis Reviews* 69 (2): 334–346.

Pautz, A. 2007. "Intentionalism and perceptual presence." *Philosophical Perspectives* 21 (1): 495–541.

Pautz, A. 2009. "What are the contents of experiences?" *Philosophical Quarterly* 59 (236): 483–507.

Pautz, A. 2013. "Do the benefits of naïve realism outweigh the costs? Comments on Fish, *Perception, Hallucination and Illusion*." *Philosophical Studies* 163: 25–36.

Pautz, A. 2010. "Do theories of consciousness rest on a mistake?" *Philosophical Issues* 20 (1): 333–367.

Pautz, A. 2020. "Representationalism about consciousness." In *The Oxford handbook of the philosophy of consciousness*, ed. U. Kriegel, 405–434. Oxford: Oxford University Press.

Perky, C. 1910. "An experimental study of imagination." *American Journal of Psychology* 21: 422–452.

Phillips, I. 2020. "Blindsight is qualitatively degraded vision." *Psychological Review* 128 (3): 558–584.

Potter, M., B. Wyble, C. Hagmann, and E. McCourt. 2014. "Detecting meaning in RSVP at 13 ms per picture." *Attention, Perception, & Psychophysics* 76 (2): 270–279.

Price, R. 2009. "Aspect-switching and visual phenomenal character." *Philosophical Quarterly* 59 (236): 508–518.

Quilty-Dunn, J. 2015. "Believing in perceiving: known illusions and the classical dual-component theory." *Pacific Philosophical Quarterly* 96: 550–575.

Rosenthal, D. 2005. *Consciousness and mind.* New York: Oxford University Press.

Russell, B. 1911. "Knowledge by acquaintance and knowledge by description." *Proceedings of the Aristotelian Society* 11 (1): 108–128.

Saad, B. 2016. "How to befriend zombies: A guide for physicalists." *Philosophical Studies* 173: 2353–2375.

Schaffer, J. 2016. "Grounding in the image of causation." *Philosophical Studies* 173 (1): 49–100.

Schaffer, J. 2017. "The ground between the gaps." *Philosophers' Imprint* 17 (11): 1–26.

Schellenberg, S. 2018. *The unity of perception: Content, consciousness, evidence.* Oxford: Oxford University Press.

Schroer, R. 2010. "Where's the beef? Phenomenal concepts as both demonstrative and substantial." *Australasian Journal of Philosophy* 88 (3): 505–522.

Schroer, R. 2014. "The Goldilocks problem of the specificity of visual phenomenal content." *Canadian Journal of Philosophy* 44 (3–4): 476–495.

Schwitzgebel, E. 2008. "The unreliability of naive introspection." *Philosophical Review* 117 (2): 245–273.

Schwitzgebel, E. 2012. "Introspection, what?" In *Introspection and consciousness*, ed. D. Smithies and D. Stoljar. Oxford: Oxford University Press.

Schwitzgebel, E. 2016. "Phenomenal consciousness, defined and defended as innocently as I can manage." *Journal of Consciousness Studies* 23 (11–12): 224–235.

Searle, J. 1983. *Intentionality: An essay in the philosophy of mind.* New York: Cambridge University Press.

Sethi, U. 2020. "Sensible over-determination." *Philosophical Quarterly* 70 (280): 588–616.

Sethi, U. 2021a. "Mind-dependence in Berkeley and the problem of perception." *Australasian Journal of Philosophy* 99 (4): 648–668.

Sethi, U. 2021b. "The varieties of instantiation." *Journal of the American Philosophical Association* 7 (3): 417–437.

Shoemaker, S. 1982. "The inverted spectrum." *Journal of Philosophy* 79: 357–381.

Sider, T. 2011. *Writing the book of the world.* Oxford: Oxford University Press.

Siegel, S. 2004. "Indiscriminability and the phenomenal." *Philosophical Studies* 120 (1–3): 91–112.

Siegel, S. 2006. "Which properties are represented in perception?" In *Perceptual experience*, ed. T. Gendler and J. Hawthorne, 481–503. New York: Oxford University Press.

Siegel, S. 2008. "The epistemic conception of hallucination." In *Disjunctivism*, eds A. Haddock and F. Macpherson, 205–224. Oxford: Oxford University Press.

Siegel, S. 2009. "The visual experience of causation." *Philosophical Quarterly* 59 (236): 519–540.

Siegel, S. 2010. *The contents of visual experience*. Oxford: Oxford University Press.

Snowdon, P. 2010. "On the what-it-is-likeness of experience." *Southern Journal of Philosophy* 48 (1): 8–27.

Soteriou, M. 2005. "The subjective view of experience and its objective commitments." *Proceedings of the Aristotelian Society* 105: 177–190.

Soteriou, M. 2016. *Disjunctivism*. New York: Routledge.

Spencer, J. 2020. "The limits of neo-Aristotelian plenitude (and its conceptual neighbors)." *Pacific Philosophical Quarterly* 101: 74–92.

Stoljar, D. 2016. "The semantics of 'What it's like' and the nature of consciousness." *Mind* 125: 1161–1198.

Sturgeon, S. 1998. "Visual experience." *Proceedings of the Aristotelian Society* 98 (1): 179–200.

Sturgeon, S. 2000. *Matters of mind*. New York, NY: Routledge.

Sturgeon, S. 2008. "Disjunctivism about visual experience." In *Disjunctivism: Perception, action, knowledge*, ed. A. Haddock and F. Macpherson, 112–143. New York: Oxford University Press.

Sundström, P. 2018. "How physicalists can—and cannot—explain the seeming 'absurdity' of physicalism." *Philosophy and Phenomenological Research* 97 (3): 681–703.

Swindle, P. 1916. "Positive after-images of long duration." *American Journal of Psychology* 27 (3): 324–333.

Tye, M. 1995. *Ten problems of consciousness: A representational theory of the phenomenal mind.* Cambridge, MA: MIT Press.

Tye, M. 2002. "Representationalism and the transparency of experience." *Noûs* 36 (1): 137–51.

Tye, M. 2009. *Consciousness revisited: Materialism without phenomenal concepts*. Cambridge, MA: MIT Press.

Tye, M. 2019. "What uninformed Mary can teach us." In *The knowledge argument*, ed. Sam Coleman, 269–279. Cambridge: Cambridge University Press.

Wilson, A. 2018. "Metaphysical causation." *Noûs* 52 (4): 723–751.

Wilson, J. 2014. "No work for a theory of grounding." *Inquiry* 57 (5–6): 535–579.

Index